D0214756

A Student's Guide to Python for Physical Modeling

Updated Edition

A Student's Guide to Python for Physical Modeling

Updated Edition

Jesse M. Kinder and Philip Nelson

Princeton University Press
Princeton and Oxford

Published by Princeton University Press, 41 William Street,
Princeton, New Jersey 08540
In the United Kingdom: Princeton University Press, 6 Oxford Street,
Woodstock, Oxfordshire OX20 1TW

press.princeton.edu

Original edition published by Princeton University Press 2015
Updated edition published 2018
Library of Congress Control Number: 2017959122

ISBN 978-0-691-18056-4
ISBN (pbk.) 978-0-691-18057-1

British Library Cataloging-in-Publication Data is available.

This book has been composed using the LaTeX typesetting system.

The publisher would like to acknowledge the authors of this volume for providing the camera-ready copy from which this book was printed.

Printed on acid-free paper. ∞

Printed in the United States of America

1 3 5 7 9 10 8 6 4 2

The front cover shows a Mandelbrot set, an image that you will be able to generate for yourself after you work through this book.

For Oliver Arthur Nelson – PN

Contents

Let's Go

Why teach yourself Python, and why do it this way

Learning to program a computer can change your perspective. At first, it feels like you are struggling along and picking up a couple of neat tricks here and there, but after a while, you start to realize that *you* can make the computer do almost *anything*. You can add in the effects of friction and air resistance that your physics professor is always telling you to ignore, you can make your own predator–prey simulations to study population models, you can create your own fractals, you can look for correlations in the stock market—the list is endless.

In order to communicate with a computer, you must first learn a language it understands. Python is an excellent choice because it is easy to get started and its structure is very natural—at least compared to some other computer languages. Soon, you will find yourself spending most of your time thinking about how to solve a problem, rather than how to explain your calculation to a computer.

Whatever your motivation for learning Python, you may wonder whether it's really necessary to wade through everything in this book. Bear with us. We are working scientists, and we have used our experience to prepare you to start exploring and learning on your own as efficiently as possible. Spend a few hours trying everything we recommend, in the order we recommend it. This will save time in the long run. We have eliminated everything you don't need at the outset. What remains is a set of basic knowledge and skills that you will almost certainly find useful someday.

How to use this tutorial

Here are a few ideas about how you might teach yourself Python using this book.

- Many code samples that appear in this document, as well as errata, updates, data sets, and more are available via `press.princeton.edu/titles/11349.html`.
- After the first few pages, you'll need to work in front of a computer that is running Python. (Don't worry—we'll tell you how to set it up.) On that same computer, you'll probably want to have open a text document named `code_samples.txt`, which is available via the website above.
- Next to the computer you may have a hard copy of this book, or the eBook on a tablet or other device. Alternatively, the eBook can be open on the same computer that runs Python.
- This book will frequently ask you to try things. Some of those things involve snippets of code given in the text. You can copy and paste code from `code_samples.txt` into your Python session, see what happens, then modify and play with it.
- You can also access the snippets interactively. A page with links to individual code samples is available via the website above. You can copy and paste code from the web page into Python.
- A few sections and footnotes are flagged with this "Track 2" symbol: $\boxed{T_2}$. These are more advanced and can be skipped on a first reading.

And now … Let's go.

CHAPTER 1

Getting Started with Python

The Analytical Engine weaves algebraical patterns, just as the Jacquard loom weaves flowers and leaves.
—Ada, Countess of Lovelace, 1815–1853

1.1 ALGORITHMS AND ALGORITHMIC THINKING

The goal of this tutorial is to get you started in computational science using the computer language Python. Python is open-source software. You can download, install, and use it anywhere. Many good introductions exist, and more are written every year. *This* one is distinguished mainly by the fact that it focuses on skills useful for solving problems in physical modeling.

Modeling a physical system can be a complicated task. Let's take a look at how we can use the powerful processors inside your computer to help.

1.1.1 Algorithmic thinking

Suppose that you need to instruct a friend how to back your car out of your driveway. Your friend has never driven a car, but it's an emergency, and your only communication channel is a phone conversation before the operation begins.

You need to break the required task down into small, explicit steps that your friend understands and can execute in sequence. For example, you might provide your friend the following set of instructions:

```
1 Put the key in the ignition.
2 Turn the key until the car starts, then let go.
3 Push the button on the shift lever and move it to "Reverse."
4 ...
```

Unfortunately, for many cars this "code" won't work, even if your friend understands each instruction: It contains a **bug**. Before step 3, many cars require that the driver

```
Press down the left pedal.
```

Also, the shifter may be marked `R` instead of `Reverse`. It is difficult at first to get used to the high degree of precision required when composing instructions like these.

Because you are giving the instructions in advance (your friend has no mobile phone), it's also wise to allow for contingencies:

```
If a crunching sound is heard, press down on the left pedal ...
```

Breaking the steps of a long operation down into small, explicit substeps and anticipating contingencies are the beginning of *algorithmic thinking.*

If your friend has had a lot of experience watching people drive cars, then the instructions above may be sufficient. But a friend from Mars—or a robot—would need much more detail. For example, the first two steps may need to be expanded to something like

```
Grab the wide end of the key.
Insert the pointed end of the key into the slot on the lower right side
        of the steering column.
Rotate the key about its long axis in the clockwise direction
        (when viewed from the wide end toward the pointed end).
...
```

These two sets of instructions illustrate the difference between low-level and high-level languages for communicating with a computer. A *low-level* computer program is similar to the second set of explicit instructions, written in a language that a machine can understand.[1] A *high-level* system understands many common tasks, and therefore can be programmed in a more condensed style, like the first set of instructions above. Python is a high-level language. It includes commands for common operations in mathematical calculations, processing text, and manipulating files. In addition, Python can access many *standard libraries*, which are collections of programs that perform advanced functions such as data visualization and image processing.

Python also comes with a *command line interpreter*—a program that executes Python commands as you type them. Thus, with Python, you can save instructions in a file and run them later, or you can type commands and execute them immediately. In contrast, many other programming languages used in scientific computing, like C, C++, or FORTRAN, require you to *compile* your programs before you can *execute* them. A separate program called a compiler translates your code into a low-level language. You then run the resulting compiled program to execute (carry out) your algorithm. With Python, it is comparatively easy to quickly write, run, and debug programs. (It still takes patience and practice, though.)

A command line interpreter combined with standard libraries and programs you write yourself provides a convenient and powerful scientific computing platform.

1.1.2 States

You have probably studied multistep mathematical proofs, perhaps long ago in geometry class. The goal of such a narrative is to establish the truth of a desired conclusion by sequentially appealing to given information and a formal system. Thus, each statement's truth, although not evident in isolation, is supposed to be straightforward in light of the preceding statements. The reader's "state" (list of propositions known to be true) changes while reading through the proof. At the end of the proof, there is an unbroken chain of logical deductions that lead from the axioms and assumptions to the result.

An **algorithm** has a different goal. It is a chain of instructions, each of which describes a simple operation, that accomplishes a complex task. The chain may involve a lot of repetition, so you won't want to supervise the execution of every step. Instead, you specify all the steps in advance, then stand back while your electronic assistant performs them rapidly. There may also be contingencies that cannot be known in advance. (`If a crunching sound is heard, ...`)

[1] Machine code and assembly language are low-level programming languages.

In an algorithm, the *computer* has a state that is constantly being modified. For example, it has many memory cells, whose contents may change during the course of an operation. Your goal might be to arrange for one or more of these cells to contain the result of some complex calculation once the algorithm has finished running. You may also want a particular graphical image to appear.

1.1.3 What does `a = a + 1` mean?

To get a computer to execute your algorithm, you must first express it in a programming language. The commands used in computer programming can be confusing at first, especially when they contradict standard mathematical usage. For example, many programming languages (including Python) accept statements such as these:

```
1  a = 100
2  a = a + 1
```

In mathematics, this makes no sense. The second line is an assertion that is always false; equivalently, it is an equation with no solution. To Python, however, "=" is not a test of equality, but an instruction to be executed. These lines have roughly the following meaning:[2]

1. Assign the name `a` (a **variable**) to an integer object with the value 100.
2. Extract the value of the object named `a`. Calculate the sum of that value and 1. Assign the name `a` to the result, and *discard* whatever was previously stored under the name `a`.

In other words, the equals sign instructs Python to change its *state*. In contrast, mathematical notation uses the equals sign to create a proposition, which may be true or false. Note, too, that Python treats the left and right sides of the command `x=y` differently, whereas in math the equals sign is symmetric. For example, Python will give an error message if you say something like `b+1=a`; the left side of an assignment must be a name that can be assigned to the result of evaluating the right side.

We do often wish to check whether a variable has a particular value. To avoid ambiguity between assignment and testing for equality, Python and many other computing languages use a double equals sign for the latter:

```
1  a = 1
2  a == 0
3  b = (a == 1)
```

The above code again creates a variable `a` and assigns it to a numerical value. Then it compares this numerical value with 0. Finally, it creates a second variable `b`, and assigns it a logical value (**`True`** or **`False`**) after performing another comparison. That value can be used in contingent code, as we'll see later.

Do not use = (assignment) when == (test for equality) is required.

This is a common mistake for beginning programmers. You can get mysterious results if you make this error, because both = and == are legitimate Python syntax. In any particular situation, however, only one of them is what you want.

---

[2] T2 Appendix E gives more precise information about the handling of assignment statements.

1.1.4 Symbolic versus numerical

In math, it's perfectly reasonable to start a derivation with "Let $b = a^2 - a$," even if the reader doesn't yet know the value of a. This statement defines b in terms of a, whatever the value of a may be.

If you launch Python and immediately give the equivalent statement, `b=a**2-a`, the result is an error message.[3] Every time you hit `<Return/Enter>`, Python tries to compute values for every assignment statement. If the variable `a` has not been assigned a value yet, evaluation fails, and Python complains. Other computer math packages can accept such input, keep track of the symbolic relationship, and evaluate it later, but basic Python does not.[4]

In math, it's also understood that a definition like "Let $b = a^2 - a$" will persist unchanged throughout the discussion. If we say, "In the case $a = 1, \ldots$" then the reader knows that b equals zero; if later we say, "In the case $a = 2, \ldots$" then we need not reiterate the definition of b for the reader to know that this symbol now represents the value $2^2 - 2 = 2$.

In contrast, a numerical system like Python *forgets* any relation between `b` and `a` after executing the assignment `b=a**2-a`. All that it remembers is the *value* now assigned to `b`. If we later change the value of `a`, the value of `b` will *not* change.[5]

Changing symbolic relationships in the middle of a proof is generally not a good idea. However, in Python, if we say `b=a**2-a`, nothing stops us from later saying `b=2**a`. The second assignment updates Python's state by discarding the value calculated in the first assignment statement and replacing it with the newly computed value.

1.2 LAUNCH PYTHON

Rather than reading about what happens when you type some command, try out the commands for yourself. Appendix A describes how to install and launch Python. From now on, you should have Python running as you read: Try every snippet of code and observe what Python does in response. For example, this tutorial won't show you any graphics or output. You must generate these yourself as you work through the examples.

> *Reading this tutorial won't teach you Python. You can teach* ***yourself*** *Python by working through all the examples and exercises here, and then using what you've learned on your own problems.*

Set yourself little challenges and test them out. ("What would happen if ...?" "How could I accomplish...?") Python is not some expensive piece of lab apparatus that could break or explode if you type something wrong! Just try things. This strategy is not only more fun than passively accumulating facts—it is also far more effective.

A complete Python programming environment has many components. See Table 1.1 for a brief description of the ones that we'll be discussing. Be aware that we use "Python" loosely in this guide. In addition to the language itself, Python may refer to a *Python interpreter*, which is a computer

---

[3] The notation `**` denotes exponentiation. See Section 1.4.2.

[4] The SymPy library makes symbolic calculations possible in Python. See Section 8.4.1.

[5] In math, the statement $b = a^2 - a$ essentially defines b as a *function* of a. We can certainly do that in Python by defining a function that returns the value of $a^2 - a$ and assigning that function the name `b` (see Section 6.1), but this is *not* what "=" does.

Python	A computer programming language. A way to describe algorithms to a computer.
IPython	A Python *interpreter:* A computer application that provides a convenient, interactive mode for executing Python commands and programs.
Spyder	An *integrated development environment* (IDE): A computer application that includes IPython, a tool to inspect variables, a text editor for writing and debugging programs, and more.
Jupyter	A notebook-style interface for Python.
NumPy	A standard library that provides numerical arrays and mathematical functions.
PyPlot	A standard library that provides visualization tools.
SciPy	A standard library that provides scientific computing tools.
Anaconda	A *distribution*: A single download that includes all of the above and provides access to many additional libraries for special purposes. It also includes a *package manager* that helps you to keep everything up to date.

Table 1.1: Elements of the Python environment described in this tutorial.

application that accepts commands and performs the steps described in a program. Python may also refer to the language together with common libraries.

Most of the code that follows will run with any Python distribution. However, since we cannot provide instructions for every available version of Python and every integrated development environment (IDE), we have chosen the following particular setup:

- The Anaconda distribution of Python 3, available at `anaconda.com`.
  Many scientists instead use an earlier version of Python (such as version 2.7). Appendix D discusses the minor changes needed to adapt the codes in this tutorial for earlier versions.
- The Spyder IDE, which comes with Anaconda or can be downloaded at `pythonhosted.org/spyder/`. Any programming task can be accomplished with a different IDE—or with no IDE at all—but this tutorial will assume you are using Spyder. Other IDEs are available, such as IDLE, which comes with every distribution of Python. Browser-based Jupyter Notebooks are another popular platform.[6]

The choice of distribution is a matter of personal preference. We chose Anaconda because it is simple to install, update, and maintain, and it is free. You may find a different distribution is better suited to your needs. Enthought Canopy is another free, widely used Python distribution.

1.2.1 IPython console

Upon launch, Spyder opens a window that includes several *panes*. See Figure 1.1. There is an Editor pane on the left for editing program files (*scripts*). There are two panes on the right. The top one may contain Variable Explorer, File Explorer, and Help tabs. If necessary, click on the Variable Explorer's tab to bring it to the front. The bottom-right pane should include a tab called "IPython Console"; if necessary, click it now.[7] It provides the command line interpreter that allows you to execute Python commands interactively as you type them.

---

[6] If you prefer the notebook interface, see Appendix B to get started. Many code samples are available in notebook format via this book's blog.

[7] If no IPython console tab is present, you can open one from the menu at the top of the screen: `Consoles>Open an IPython console`.

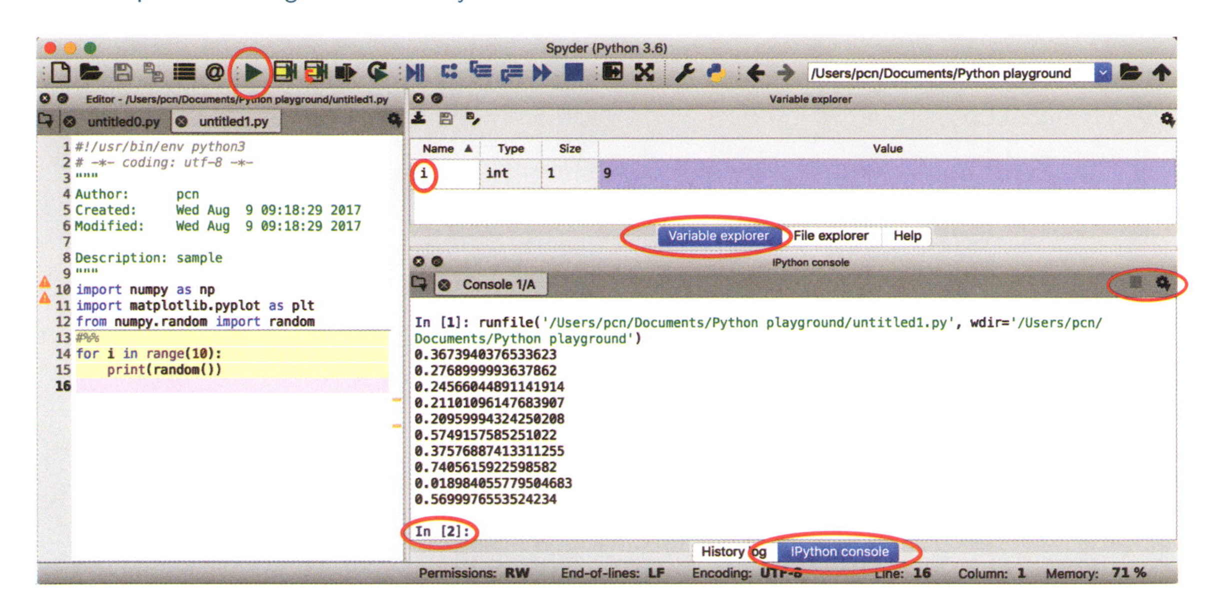

Figure 1.1: The Spyder display. Red circles have been added to emphasize (*from top to bottom*) the Run ▶ button, a variable in the Variable Explorer, the tab that brings the Variable Explorer to the front in its pane, the Stop ■ and Options ⚙ buttons, the IPython command prompt, and the IPython Console tab. Two scripts are open in the Editor; `untitled1.py` has been brought to the front by clicking its tab at the top of the Editor pane.

If your window layout gets disorganized, do not worry. It is easy to adjust. The standard format for Spyder is to open with a single window, divided into the three panes just described. Each pane can have multiple tabs. If you have unwanted windows, close them individually by clicking on their Close ⊗ buttons. You can also use the menu `View>Panes` to select panes you want to be visible and deactivate those you do not want. `View>Window layouts>Spyder Default Layout` will restore the standard layout.

Click in the IPython console. Now, things you type will show up after the *command prompt*. By default, this will be something like

```
In[1]:
```

Try typing the lines of code in Section 1.1.3, hitting `<Return/Enter>` after each line. Python responds immediately after each `<Return/Enter>`, attempting to perform whatever command you entered.[8]

Python code consists entirely of plain text.

All fonts, typefaces, and coloring in the code samples of this tutorial were added for readability. These are not things you need to worry about while entering code. Similarly, the line numbers shown on the left of code samples are there to allow us to refer to particular lines. Don't type them. Spyder will assign and show line numbers when you work in the Editor, and Python will use them to tell you where it thinks you have made errors. They are not part of the code. Note also that most blank spaces

[8] This tutorial uses the word "command" to mean any Python statement that can be executed by an interpreter. Assignments like `a=1`, function calls like `plt.plot(x,y)`, and special instructions like `%reset` are commands.

are optional, except when used for indentation. We use extra blank spaces to improve readability, but these are not required.

This tutorial uses the following color and font scheme when displaying code:

- Built-in functions and reserved words are displayed in boldface green type: `print("Hello, world!")`. You do not need to import these functions.
- Python errors and runtime exceptions are displayed in boldface red type: **`SyntaxError`**. These are also part of basic Python.
- Functions and other objects from NumPy and PyPlot are displayed in boldface black type: `np.sqrt(2)`, or `plt.plot(x,y)`. We will assume that you import NumPy and PyPlot at the beginning of each session and program you write.[9]
- Functions imported from other libraries are displayed in blue boldface type: `from scipy.special import factorial`.
- Strings are displayed in red type: `print("Hello, world!")`.
- Comments are displayed in oblique blue type: `# This is a comment.`
- Keywords in function arguments are displayed in oblique black type: `np.loadtxt('data.csv', delimiter=',')`. Keywords are not arbitrary; they must be spelled exactly as shown.
- Keystrokes are displayed within angled brackets: <Enter/Return> or <Ctrl-C>.
- Buttons you can click with a mouse are displayed in small capitals within a rectangle: [RUN ▶]. Some buttons in Spyder have icons rather than text, but hovering the mouse pointer over the button will display the text shown in this tutorial.
- Most other text is displayed in normal type.

Click on the Variable Explorer tab. Each time you enter a command and hit <Return/Enter>, the contents of this pane will reflect any changes in Python's state: Initially empty, it will display a list of your variables and a summary of their values.[10] When a variable contains many values (for example, an array), you can double-click its entry in this list to open a spreadsheet that contains all the values of the array. You can copy from this spreadsheet and paste into other applications.

At any time, you can reset Python's state by quitting and relaunching it, or by executing the command

```
%reset
```

Since you are about to delete everything that has been created in this session, you will be asked to confirm this irreversible operation.[11] Press <y> then <Return/Enter> to proceed. (Commands that begin with a % symbol are **magic commands**, that is, commands specific to the IPython interpreter. They may not work in a basic Python interpreter, or in scripts that you write. To learn more about these, type `%magic` at the IPython command prompt.)

---

[9] See Section 1.3 (page 11).

[10] Some variables will not appear. You can control which variables are excluded through the [OPTIONS ⚙] menu, in the upper-right corner of the Variable Explorer pane.

[11] If IPython does not seem to respond to `%reset`, try scrolling the IPython console up manually to see the confirm query.

Example: Use the `%reset` command, then try the following commands at the prompt. Explain everything you see happen:

```
q
q == 2
q = 2
q
q == 2
q == 3
```

Solution: Python complains about the first two lines: Initially, the symbol `q` is not associated with any object. It has no value, and so expressions involving it cannot be evaluated. Altering Python's state in the third line above changes this situation, so the last three lines do not generate errors.

Example: Now clear Python's state again. Try the following at the prompt, and explain everything that happens. (It may be useful to refer to Section 1.1.4.)

```
a = 1
a
b = a**2 - a
b
a = 2
print(a)
print(b)
b = a**2 - a
a, b
print(a, b)
```

Solution: The results from the first four lines should be clear: We assign values to the variables `a` and `b`. In the fifth line, we change the value of `a`, but because Python remembers only the *value* of `b` and not its relation to `a`, its value is unchanged until we update it explicitly in the eighth line.

When entering code at the command prompt, you may run into a confusing situation where Python seems unresponsive.

> *If a command contains an unmatched* `(`, `[`, *or* `{`, *then Python continues reading more lines, searching for the corresponding* `)`, `]`, *or* `}`.

If you cannot figure out how to match up your brackets, you can abort the command by pressing `<Esc>`.[12] You can then retype the command and proceed with your work.

The examples above illustrate an important point: An assignment statement does not display the value that it assigns to a variable. To see the value assigned to a variable in an IPython session, enter the variable name on a line by itself. Alternatively, the `print()` command can be used to display values.[13] Note that "print" does not cause anything to come out of a printer; instead, it displays the requested information in the IPython console. If you need hard copy output, you can have your program write the desired information to a plain text file (see Section 4.2.2), then print that file in the usual way.

---

[12] `<Esc>` cancels the current command in Spyder. In another IDE or interpreter, you may need to use `<Ctrl-C>` instead.

[13] In scripts that you write, Python will evaluate an expression *without* showing anything on the screen; if you want output, you must give an explicit `print()` command. Scripts will be discussed in Section 3.3.

The last two lines of the example above illustrate how to see the values of multiple objects at once. Notice that the output is not exactly the same.

You can end a command by starting a new line. Or, if you wish, you can end a command with a semicolon (`;`) and then add another command on the same line. It is also possible to make multiple assignments with a single = command. This is an alternative to using semicolons. Both of the following lines assign the same values to their respective variables:

```
a = 1; b = 2; c = 3
a, b, c = 1, 2, 3
```

Either side of the second command may be enclosed in parentheses without affecting the result.

The preceding paragraph demonstrates ways to save space and reduce typing with Python. Sometimes this is convenient, but it's best not to make too much use of this ability. You should instead try to make the *meaning* of your code as clear as possible. Human readability is worth a few extra seconds of typing or a few extra lines in a file.

In some situations, you may wish to use a very long command that doesn't fit on one line. For such cases, you can end a line with a backslash (\). Python will then continue reading the next line as part of the same command. Try this:

```
q = 1 + \
2
q
```

A single command can even stretch over multiple lines:

```
xv\
a\
l\
= 1 + \
2
```

This will create a variable `xval` assigned to the value 3. To write clear code, you should use this option sparingly.

1.2.2 Error messages

You should have encountered some error messages by now. When Python detects an error, it tells you where it encountered the error, provides a fragment of the code surrounding the statement that caused the problem, and tells you which general kind of error it detected among the many types it recognizes. For example, Python responds with a **`NameError`** whenever you try to evaluate an undefined variable. (Recall the Example on page 8.) See Appendix C for a description of common Python errors and some hints for interpreting the resulting messages.

1.2.3 Sources of help

The definitive documentation on Python is available online at `www.python.org/doc`. However, in many cases you'll find the answers you need more quickly by other means, such as asking a friend, searching the web, or visiting `stackoverflow.com`.

Suppose that you wish to evaluate the square root of 2. You type `2**0.5` and hit `<Return/Enter>`. That does the job, but Python is displaying 16 digits after the decimal point, and you only want 3.

You think there's probably a function called **round** in Python, but you are not sure how to use it or how it works. You can get help directly from Python by typing **help**(**round**) at the command prompt. You'll see that this is indeed the function you were looking for:

```
round(2**0.5, 3)
```

gives the desired result.

In Spyder, there are additional ways to get help. Type **round** at the command prompt, but do not hit <Return/Enter>. Instead hit <Cmd-I> or <Ctrl-I> (for "Information"). The information that was displayed in the IPython console when you issued the **help** command now shows up in the Help tab, and in a format that is easier to navigate and read, especially for long entries. You can also use the Help tab without entering anything at the command prompt: Try entering **pow** in the "Object" field at the top of the pane. The Help tab will provide information about an alternative to the ** operation for raising a number to a power.

In IPython, you can also follow or precede the name of a module,[14] function, or variable by a question mark to obtain help: **round?** or **?round** provides the same information as **help**(**round**) and is easier to type.

When you type **help**(...), Python will print out the information it has about the expression in parentheses if it recognizes the name. Unfortunately, Python is not as friendly if you don't know the name of the command you need. Perhaps you think there ought to be a way to take the square root of a number without using the power notation. After all, it is a pretty basic operation. Type **help**(sqrt) to see what happens when Python does not recognize the name you request.

To find out what commands are currently available to you, you can use Python's **dir**() command. This is short for "directory," and it returns a list of all the modules, functions, and variable names that have been created or imported during the current session (or since the last **%reset** command). Ask Python for help on **dir** to learn more. Nothing in the output of **dir** looks promising, but there is an item called __builtin__. This is the collection of all the functions and other objects that Python recognizes when it first starts up. It is Python's "last resort" when hunting for a function or variable.[15] To see the list of built-in functions, type

```
dir(__builtin__)
```

There is no sqrt function or anything like it. In fact, *none* of the standard mathematical functions, such as sin, cos, or exp show up!

Python cannot help you any further at this point. You now have to turn to outside resources. Good options include books about Python, search engines, friends who know more about Python than you do, and so on.

> *In the beginning, a lot of your coding time will be spent using a search engine to get help.*

The **sqrt** function we seek belongs to a library. Later we will discuss how to access libraries of useful functions that are not automatically available with Python.

Your Turn 1A

> Before proceeding, try a web search for
> how to take square roots in python

[14] Modules will be discussed in Section 1.3.

[15] Appendix E explains how Python searches for variables and other objects.

1.2.4 Good practice: Keep a log

As you work through this tutorial, you will hit many small roadblocks—and some large ones. How do you evaluate a modified Bessel function? What do you do if you want a subscript in a graph axis label? The list is endless. Every time you resolve such a puzzle (or a friend helps you), *make a note of how you did it* in a notebook or in a dedicated file somewhere on your computer. Later, looking through that log will be much easier than scanning through all the code you wrote months ago (and less irritating than asking your friend over and over).

1.3 PYTHON MODULES

We discovered that Python does not have a built-in `sqrt` function. Even your calculator has that! What good is Python? Think for a moment about how, exactly, your calculator knows how to find square roots. At some point in the past, someone came up with an algorithm for computing the square root of a number and stored it in the permanent memory of your calculator. Someone had to create a *program* to calculate square roots.

Python is a programming language. A Python interpreter understands a basic set of commands that can be combined to perform complex tasks. Python also has a large community of developers who have created entire libraries of useful functions. To gain access to these, however, you need to **import** them into your working environment.

> *Use the* **`import`** *command to access functions that do not come standard with Python.*

1.3.1 `import`

At the command prompt, type

```
import numpy
```

and hit `<Return/Enter>`. You now have access to many useful functions. You have imported the NumPy module, a collection of tools for numerical calculation using Python: "Numerical Python." (Do not capitalize its name in your code.)

To see what has been gained, type `dir(numpy)`. You will find nearly 600 new options at your disposal, and one of them is the `sqrt` function you originally sought. You can search for the function within NumPy by using the command `numpy.lookfor('sqrt')` (This will often return more than you need, but the first few lines can be quite helpful.) Now that you have imported NumPy, try

```
sqrt(2)
```

What's going on? You just imported a square root function, but Python tells you that `sqrt` is not defined! Try instead

```
numpy.sqrt(2)
```

The `sqrt` function you want "belongs" to the **numpy** module you imported. Even after importing, you still have to tell Python where to find it before you can use it.

> *After you have imported a module, you can call its functions by giving the module name, a period, and then the name of the desired function.*

1.3.2 from ... import

There is another way to import functions. For example, you may wish access to all of the functions in NumPy without having to type the "**numpy.**" prefix before them. Try this:

```
from numpy import *
sqrt(2)
```

This is convenient, but it can lead to trouble when you want to use two different modules simultaneously. There is a module called **math** that also has a sqrt function. If you import all of the functions from **math** and **numpy**, which one gets called when you type sqrt(2)? (This is important when you are working with arrays of numbers.) To keep things straight, it is usually best to avoid the "**from** module **import** *" command. Instead, import a module and explicitly call **numpy.sqrt** or **math.sqrt** as appropriate. However, there is a middle ground. You can give a module any *nickname* you want. Try this:

```
import numpy as np
np.sqrt(2)
```

Now we can save typing and still avoid confusion when functions from different modules have the same name.

There may be times when you only want a specific function, not a whole library of functions. You can ask for specific functions by name:

```
from numpy import sqrt, exp
sqrt(2)
exp(3)
```

We now have just two functions from the NumPy module, which can be accessed without the "**numpy.**" prefix. Notice the similarity with the "**from numpy import** *" command. The asterisk is a "wildcard" that tells the import command to grab everything.

There is one more useful variant of importing that allows you to give the function you import a custom nickname:

```
from numpy.random import random as rng
rng()
```

We now have a random number generator with the convenient name rng.

This example also illustrates a module within a module: **numpy** contains a module called **numpy.random**, which in turn contains the function **numpy.random.random**. When we typed **import numpy**, we imported many such subsidiary modules. Instead, we can import just one function by using **from** and providing a precise specification of the function we want, where to find it, and what to call it.

1.3.3 NumPy and PyPlot

The two modules we will use most often are called NumPy and PyPlot. NumPy provides the numerical tools we need to generate and analyze data, and PyPlot provides the tools we need to visualize data. PyPlot is a subset of the much larger Matplotlib library. From now on, we will assume that you have issued the following commands:

```
import numpy as np
import matplotlib.pyplot as plt
```

This can also be accomplished with the single command

```
import numpy as np, matplotlib.pyplot as plt
```

You should execute these commands at the start of every session. You should also add these lines at the beginning of any scripts that you write. You will also need to reimport both modules each time you use the `%reset` command.

Give the `%reset` command, then try importing these modules now. Explore some of the functions available from NumPy and PyPlot. You can get information about any of them by using `help()` or any of the procedures described in Section 1.2.3. You will probably find the NumPy help files considerably more informative than those for the built-in Python functions. They often include examples that you can try at the command prompt.

Now that we have these collections of tools at our disposal, let's see what we can do with them.

1.4 PYTHON EXPRESSIONS

The Python language has a **syntax**—a set of rules for constructing expressions and statements. In this section, we will look at some simple expressions to get an idea of how to communicate with Python. The basic building blocks of expressions are literals, variable names, operators, and functions.

1.4.1 Numbers

You can enter explicit numerical values (numeric **literals**) in various ways:

- `123` and `1.23` mean what you might expect. When entering a large number, however, don't separate groups of digits by commas. (Don't type `1,000,000` if you mean a million.)
- `2.3e5` is convenient shorthand for $2.3 \cdot 10^5$.
- `2+3j` represents the complex number $2 + 3\sqrt{-1}$. (Engineers may find the name j for $\sqrt{-1}$ familiar; mathematicians and physicists will have to adjust to Python's convention.)

Python stores numbers internally in several different formats. However, it will usually convert from one type to another when necessary. Beginners generally don't need to consider this. Just be aware that Python sometimes requires an integer. Even if a value has no fractional part, Python may not interpret it as an integer (for example, `a=1.0`). If you need to force a value to be an integer (for example, when indicating an entry in a list), you can use the function `int`.

1.4.2 Arithmetic operations and predefined functions

Python includes basic arithmetic operators, for example, `+`, `-`, `*` (multiplication), `/` (division), and `**` (exponentiation).

Python uses two asterisks, `**`, to denote raising a number to a power.

For example, `a**2` means "a squared." (The notation `a^2` is used by some other math software but means something quite different to Python.)

Unlike standard mathematics notation, you may not omit multiplication signs. Try typing

```
(2)(3)
a = 2; a(3)
3a
3 a
```

Each of these commands produces an error message. None, however, generates a message like, "`You forgot a '*'!`" Python used its evaluation rules, and these expressions didn't make sense. Python doesn't know what you were trying to express, so it can't tell you exactly what is wrong. *Study these error messages*; you'll probably see them again. See Appendix C for a description of these and other common errors.

Arithmetic operations have the usual precedence (ordering).

You can use parentheses to override operator precedence.

Unlike math textbooks, Python recognizes only parentheses (round brackets) for ordering operations. Square and curly brackets are reserved for other purposes. We have already seen that parentheses can also have another meaning (enclosing the arguments of a function). Yet another meaning will appear later: specifying a tuple. Python uses context to figure out which meaning to use.

For example, if you want to use the number $\frac{1}{2\pi}$, you might type `1/2*np.pi`. (Basic Python does not know the value of π, but NumPy does.) Try it. What goes wrong, and why? You can fix the expression by inserting parentheses. Later we'll meet other kinds of operators such as comparisons and logical operations. They, too, have a precedence ordering, which you may not wish to memorize. Instead, use parentheses liberally to specify precisely what you mean.

To get used to Python arithmetic operations, figure out what famous math problem these lines solve, and check that Python got it right:

```
a, b, c = 1, -1, -2
(-b + np.sqrt(b**2 - 4*a*c))/(2*a)
```

Recall that **`np.sqrt`** is the name of a *function* that Python does not recognize when it launches, but that becomes available once we import the NumPy module. When Python encounters the expression in the second line, it does the following:

1. Evaluate the **argument** of the **`np.sqrt`** function—that is, everything inside the pair of parentheses that follows the function name—by substituting values for variables and evaluating arithmetic operations. (The argument may itself contain functions.)
2. Interrupt evaluation of the expression and execute a piece of code named **`np.sqrt`**, handing that code the result found in step **1**.
3. Substitute the value returned by **`np.sqrt`** into the expression.
4. Finish evaluating the expression as usual.

How do you know what functions are available for you? See Section 1.2.3 above: Type **`dir`**`(`**`np`**`)` and **`dir`**`(__builtin__)` at the IPython console prompt.

A few symbols in Python and NumPy are predefined. These do not require any arguments or parentheses. Try **`np.pi`** (the constant π), **`np.e`** (the base of natural logarithms e), and **`1j`** (the constant $\sqrt{-1}$). NumPy also provides the standard trig functions, but be alert when using them:

The trig functions **`np.sin`**, **`np.cos`**, *and* **`np.tan`** *all treat their arguments as angles expressed in radians.*

1.4.3 Good practice: Variable names

Note that Python offers you no protection against accidentally changing the value of a symbol: If you say **np.pi**=22/7, then until you change it or reset Python, **np.pi** will have that value. It is even possible to create a variable whose name supplants a built-in function, for example, **round**=3.[16] This illustrates another good reason for using the "**import numpy as np**" command instead of the "**from numpy import** *" command: You are quite unlikely to use the "**np.**" prefix and name your *own* variables **np.pi** or **np.e**. Those variables retain their standard values no matter how you define pi and e.

When your code gets long, you may inadvertently reuse variable names. If you assign a variable with a generic name like x in the beginning, you may later choose the same name for some completely different purpose. Later still, you will want the original x, having forgotten about the new one. Python will have overwritten the value you wanted, and puzzling behavior will ensue. You have a **name collision**.

It's good practice to use longer, more meaningful names for variables. They take longer to type, but they help avoid name collisions and make your code easier to read. Perhaps the first variable you were planning to call x could instead be called index, because it indexes a list. Perhaps the second variable you were planning to call x could logically be called total. Later, when you ask for index, there will be no problem.

Keep in mind, however, that "meaningful" in this context implies "meaningful to a human reader." Python itself pays no attention to the meaning of your variable names; for example, naming a variable filename will not tell Python how to use that variable.

Variable names are case sensitive, and most predefined names are lowercase. Thus, you can avoid some name collisions by including capital letters in variable or function names you define.

Blank spaces and periods are not allowed in variable names. Some coders use capitalization in the middle of variable names ("camel humps") to denote word boundaries—for example, whichItem. Others use the underscore, as in which_item. Variable names may also contain digits (myCount2), but they must start with a letter.

Some variable names are forbidden. Python won't let you name variables **if**, **for**, **lambda**, or a handful of other **reserved words**. You can find them with a web search for python reserved words.

1.4.4 More about functions

You may be accustomed to thinking of a function, for example, square root, as a machine that eats one number (its argument) and spits out another number (its result). Some Python functions do have this character, but Python has a much broader notion of function. Here are some illustrations. (Some involve functions that we have not seen yet.)

- A function may have a single argument, multiple arguments separated by commas, or no arguments at all.
- A function may allow a *variable* number of arguments, and behave differently depending on how many you supply. For example, we will see functions that allow you to specify options by using *keyword arguments*. Each function's help text will describe the allowed ways of using it.
- A function may also *return* more than one value. The number of values returned can even vary

[16] This can be undone by deleting your version of **round**: Type **del**(**round**). Python will revert to its built-in definition.

depending on the arguments you supply. You can capture the returned values by using a special kind of assignment statement.[17]

- A function may change your computer's state in ways other than by returning a result. For example, **plt.savefig** saves a plot to a file on your computer's hard drive. Other possible side effects include writing text into the IPython console: **print**('hello').

If you use a function name without any parentheses, you are referring to the function instead of evaluating it. In mathematics, f is a function; $f(2)$ is the value of the function when its argument is 2. Type **np.sqrt** with no parentheses at the IPython command line to see how Python handles function names.

> *When evaluating a function, always include parentheses—even if there are no arguments.*

If a function accepts two or more arguments, how does it know which is which? In mathematical notation, the *order* of arguments conveys this information. For example, if we define $f(x, y) = x\,\mathrm{e}^{-y}$, then later $f(2, 6)$ means $2 \cdot \mathrm{e}^{-6}$: The first given value (2) gets substituted for the first named variable in the definition (x), and so on. This **positional argument** scheme is also the standard one used by Python. But when a function accepts many arguments, relying on order can be annoying and prone to error. For this reason, Python has an alternative approach called **keyword arguments**. For example,

```
f(y=6, x=2)
```

instructs Python to execute a function named `f`, initializing a variable named `y` with the value 6 and another named `x` with the value 2. You need not adhere to any particular order in giving keyword arguments. (However, keyword arguments must follow all positional arguments and you must use their correct names, which you can learn from the function's documentation.) Many functions will let you omit specifying values for some or all of their keyword arguments; if you omit them, the function supplies default values. Keyword arguments will be discussed further in Section 6.1.3.

You now know enough Python to start doing simple calculations. Try the examples from this chapter and play around on your own. In the next chapter, we will explore how to write simple programs in Python.

[17] Section 6.3.1 (page 76) discusses the values returned by functions in more detail. T2 More precisely, a Python function always returns a single object. However, this object may be a tuple that contains several items.

CHAPTER 2

Organizing Data

Heisenbug: *A computer bug that disappears or alters its characteristics when an attempt is made to study it.*
— Wikipedia

Much of the power of computation comes from the ability to do repetitive tasks extremely rapidly. You need to understand how to formulate instructions for this sort of task, so that your electronic assistant can do all the steps without your supervision. An important part of the process is organizing the data. This chapter describes several Python *data structures* that are useful in scientific computing.

2.1 OBJECTS AND THEIR METHODS

In Python, everything is an **object**. An object is a combination of data and functions. Even simple things like integers are objects in Python. (Type `dir(1)` to see all the data and functions associated with what you might have thought was "just a number.") When processing an assignment statement, Python attaches, or **binds**, a name (a symbol like `x` or `filename`) to an object.[1] Later, you can refer to the object by this name, or you can reassign the name to a different object. Names used in this way are also called **variables**.

Let's look at a few examples to see how this works.

When you type `i=5280`, Python begins by evaluating the right-hand side of the assignment statement. The only thing it finds is a numeric **literal**, that is, an expression whose value is the expression itself. (Thus, `1+1` is an expression, but it is not a literal because its value is `2`. In contrast, `5280` is a literal because its value is 5280.) Continuing with the assignment, Python creates an integer (object of type `int`) to store the number 5280. To complete the assignment, Python binds the name `i` to the new object. If no variable with the name `i` exists, then Python will create one. If `i` already exists, Python reassigns it to the new integer object.

When you type `f=2.5`, or `f=2.5e30`, Python creates a different type of object, called a `float`, which is short for **floating-point number**. As in scientific notation, the decimal point can "float" to give a constant number of significant figures, regardless of how large or small the number is.

Similarly, typing `s='Hello'` creates a `str` object, whose value is the string literal `'Hello'`, then binds the name `s` to that object. (Section 2.3 will discuss strings.) It does not matter if `s` was previously bound to a `float` or some other sort of object. Types belong to objects, not to the variables that point to them.

Writing `L=[1,2,3]` creates a `list` object whose value is a sequence of three `int` objects. The value of an `int` object is restricted to be an integer. (Section 2.2 will discuss lists.)

[1] T2 Appendix E describes in greater detail how Python handles assignment statements.

There is more to an object than just its value. In general, objects consist of both **attributes** (data) and **methods** (functions). The value of an object is one of its attributes. A method of an object is a specialized function that can act on its attributes. A method may also accept additional arguments. To access a method, you have to provide the name of the object, followed by a period, the name of the method, and a pair of parentheses enclosing any arguments. Try the following with the `float` object `f`, the `str` object `s`, and the `list` object `L` introduced above:

1. `f.is_integer()` : Every `float` object has a method called **`is_integer`** that determines whether its value is equivalent to an integer. The method returns **`True`** if the fractional part of the value is zero. Because the value of `f` is 2.5, the method returns the value **`False`**. The function `f.is_integer` does not require any additional arguments, but we must nevertheless place an empty pair of parentheses after its name.
2. `s.swapcase()` : Every `str` object has a method called **`swapcase`** that returns a new string whose value is the original string with the case of every letter reversed (upper↔lower).
3. `L.reverse()`: Every `list` object has a method called **`reverse`** that does not return a value but does modify the value of `L`. The output of `print(L)` will show that the order of the list has been reversed.
4. `L.pop(0)` : Every list also has a method called **`pop`** that will return the item at a specified location and remove it from the list. Thus, this method returns a value *and* modifies the object's data. When called with no argument, `L.pop()` will remove the last item in the list.

These examples illustrate that every object possesses a suite of attributes and methods. Even literals have the standard methods appropriate to their type: Try `'Hello'.swapcase()`, or `(5.0).is_integer()`. You can see all the methods associated with an object by using the **`dir`** function.

A method can use its parent object's data (examples **1**–**4** above). It can modify that data (examples **3, 4**) or not (examples **1, 2**). It can return a value (examples **1, 2, 4**) or not (example **3**). It can accept additional arguments in parentheses (example **4**) or not (examples **1**–**3**). Each method's documentation describes its behavior.

There are some objects in Python whose values are fixed once the object is created. They are called **immutable objects**.

The methods of an immutable object do not change its value.

String and numeric objects are immutable, which is why examples **1** and **2** above do not change their values.[2] Lists are **mutable**: Their methods can modify their data. Examples **3** and **4** illustrated this. Shortly, we'll introduce NumPy arrays, which are also mutable.

Try creating all the example objects mentioned in this section and see how they appear in Spyder's Variable Explorer pane. Experiment with some of their methods.

Many objects have attributes that describe properties of the object other than its value. These are stored as part of the object and require no computation. Python just looks up the attribute and returns its value (see Section 2.2.2). The syntax for accessing an attribute is `object.attribute`. For example, try `q=1+3j` followed by `q.imag`. No parentheses are needed because an attribute is not a function. The expression starts with the name of the broader thing (here `q`), followed by a period and the name of a specific thing (here `imag`).

---

[2] The method **`swapcase`** returns a modified copy of the original string; it does not affect the original string.

You can create your own objects, and give them whatever attributes and methods you like, but that goes beyond the scope of this tutorial. To learn more, try a web search for "`python classes`."

2.2 LISTS, TUPLES, AND ARRAYS

Much of the power of computer programming comes when we handle numbers in batches. A batch of numbers could represent a single mathematical object such as a force vector, or it could be a set of points at which you wish to evaluate a function. To process a batch of numbers, we first need to collect them into a single data structure. The most convenient Python data structure for our purposes is the NumPy array, described below. Lists and tuples are also useful.

2.2.1 Creating a list or tuple

Python comes with a built-in `list` object type. Any set of objects enclosed by square brackets and separated by commas creates a list:[3]

```
L = [1, 'a', max, 3 + 4j, 'Hello, world!']
```

A **tuple** is similar to a list, but immutable. To create a `tuple` object, type `t=(2,3,4)`. Note the use of round parentheses, `()`, in contrast to the square brackets, `[]`, of a list. Python knows you want a tuple because it scans the text between the parentheses and finds a comma. Without any commas, parentheses are interpreted as specifying the order of operations.[4] Like a list, a tuple consists of an ordered sequence of objects. Unlike a list, however, a tuple is immutable. Its elements cannot be reassigned, nor can their order be modified. We will use tuples to specify the shape of an array in Section 2.2.2 and later sections. In addition, functions may return a tuple containing several objects.

2.2.2 NumPy arrays

A list is a sequence (ordered collection) of objects, endowed with methods that can perform certain operations on its contents like searching and counting. You can do numerical calculations with lists; however, a `list` object is generally *not* the most convenient data type for such work. What we usually want is a *numerical array*—a grid of numbers, all of the same type. The NumPy module provides a class of array objects ideally suited to our needs. Knowing that all of the elements in an array are numbers of the same type allows NumPy to do efficient calculations on an entire array. Because Python lists can contain any mix of object types, processing their elements is far less efficient.

A Python list is ***not*** *the same as a NumPy array.*
We will almost always use NumPy arrays.

You can create a one-dimensional array by using the command

```
a = np.zeros(4)
```

The function **`np.zeros`** requires one argument that describes the shape of the array; in this case, the integer 4 specifies an array with four elements. It sets all the entries to zero.

In mathematics, we often need arrays of numbers with two or more dimensions. Try

---

[3] A list can contain just one object, but note that `2.71`, a `float` object, is not the same thing as `[2.71]`, a list containing one `float` object. A list can even contain *no* objects: `L=[]`.

[4] If you want a tuple with just one element, write `(0,)`.

```
a = np.zeros( (3, 5) )
```

and see what you get. Here **np.zeros** again accepts a single argument. In this case, it is the tuple `(3,5)`. The function creates an array with 15 entries, all equal to zero, arranged in three rows and five columns. The name `a` now refers to a `ndarray` object (an "n-dimensional NumPy array") that conforms to the mathematical notion of a 3×5 matrix.

Your Turn 2A

> Create `a` by using the preceding code. Then, find it in the Variable Explorer. Double-click its value to look inside the array. Repeat with the function **np.ones** in place of **np.zeros** and see what you get. Finally, try **np.eye**`(3)`.

A *row vector* is a special case of a two-dimensional array with just one row: **np.zeros**`((1,5))`. Similarly, a *column vector* is a two-dimensional array with just one column: **np.zeros**`((3,1))`. Python will give you exactly what you ask for, so be aware of what you are requesting:

> *If you want a column vector of N zeros, use* **np.zeros**`((N,1))`*, not* **np.zeros**`(N)`. *If you want a row vector of N zeros, use* **np.zeros**`((1,N))`.

The functions **np.ones** and **np.random.random** can also create two-dimensional arrays, using the same syntax.

Python can report how big an array is. After setting up `a`, use the command **np.size**`(a)` and see what you get. Also try **np.shape**`(a)`. This information is also displayed in the Variable Explorer.

A NumPy array is an object with its own attributes. For example, it can report its own size and shape (`a.`**shape** and `a.`**size**). A NumPy array also has methods for acting on its data. Try `a.`**sum**`()`, `a.`**mean**`()`, and `a.`**std**`()`. (Of course, these methods are more useful when the array contains something other than zeros!)

Compare the output of **np.zeros**`(3)`, **np.zeros**`((1,3))`, and **np.zeros**`((3,1))`. Note, in particular, the shape of each. NumPy regards the first as neither a row nor a column vector. It is a one-dimensional array—an array with only one index. In some cases, NumPy will treat all three types of arrays the same. However, in certain matrix and vector operations, the shape of the array is important. Be certain you have arrays of the shape you need when performing these operations. You can reshape a one-dimensional array into a column vector or row vector if necessary. (See Section 2.2.9.)

Python can also handle arrays with more than two dimensions. Try

```
A = np.zeros( (2, 3, 4) )
B = np.ones( (2, 3, 4, 3) )
```

and inspect the resulting arrays.

Don't confuse the concept of "three-dimensional array" with that of "three-dimensional vector." The variable `A` defined earlier is a three-dimensional array; it could represent grid points filling a volume of space, each holding a single quantity (a *scalar*). In contrast, a three-dimensional vector is just a set of three numbers (the components of the vector); it could be represented by an array like **np.ones**`(3)` or **np.ones**`((3,1))`.[5]

---

[5] T2 A three-dimensional grid of three-dimensional vectors can be represented by a *four*-dimensional NumPy array!

2.2.3 Filling an array with values

Perhaps you'd like to set up an array with more interesting values. The command

```
a = [2.71, 3.14, 3000]
```

creates a Python list—*not* a NumPy array—with the specified values. It contains the values we want, but we cannot perform the operations we would like with the list. To create a NumPy array from these values, we can call the function **np.array** on the list:

```
a = np.array( [2.71, 3.14, 3000] )
```

Making each element of the list its own list—that is, enclosing each entry in square brackets—returns a column vector, a two-dimensional array with three rows and one column:

```
a = np.array( [ [2.71], [3.14], [3000] ] )
```

Your Turn 2B

Try

```
a = np.array( [ [2, 3, 5], [7, 11, 13] ] )
```

and explain the result.

Usually you don't want to specify each entry in an array explicitly. For example, we frequently want to create a NumPy array of evenly spaced values over some range. NumPy provides two useful functions to do this: **np.arange** and **np.linspace**.

The function **np.arange**(M,N) creates a one-dimensional array with M as its first entry, M+1 as the second, and so on, stopping just *before* it reaches a value that equals or exceeds N.

Example: Try the following commands. Inspect and describe the resulting arrays.

```
a = np.arange(1, 10)
b = np.arange(5)
c = np.arange(2.1, 5.4, 0.1)
```

The last command creates a series, but each entry is greater than its predecessor by 0.1, not by the default increment of 1.

> *The* **arange** *syntax is* **np.arange**(start,end,increment)*. The increment and starting value are optional. The default start value is 0; the default increment is 1. The end value* ***is not included*** *in the array.*

Type **help**(**np.arange**) at the command prompt to learn more.

The function **np.linspace**(A,B,N) does a similar job. It creates a one-dimensional array with exactly N evenly spaced entries. The first entry equals A. Unlike the **np.arange** function, however, the last entry equals B exactly. You don't get to specify the spacing; Python determines what is needed to cover the range from A to B with N equally spaced points.

> *The* **linspace** *syntax is* **np.linspace**(start,end,num_points)*. The end value* ***is included*** *in the array.*

Your Turn 2C

a. Try

```
a = np.arange(0, 10, 2)
b = np.linspace(0, 10, 6)
```

and explain the results. See if you can use **np.arange** and **np.linspace** to create identical arrays. [*Hint:* What happens when you add the increment to the end value in **np.arange**?]
b. Now try

```
a = np.arange(0, 10, 1.5)
b = np.linspace(0, 10, 7)
```

Describe how a and b differ. Explain why, and decide which form you should use if you want to evaluate a function over the range 0 to 10.

When creating a series to compute the values of a function over a range, **np.linspace** is the appropriate function. It allows you to explicitly choose the start and end of the range (and the number of points in the series). However, when the exact spacing between points is important, use **np.arange** instead.

If you want to control both the end point and the spacing, the following construction will do the job:

```
x_min = 0
x_max = 10
dx = 0.1
x_array = np.arange(x_min, x_max + dx, dx)
```

Python also has a built-in function called **range**, but it does not create a numerical array of values. Instead, it creates an object that returns a sequence of values, one at a time. This makes **range** useful for loops, as described in Section 3.1, but it should not be used in place of an array.

2.2.4 Concatenation of arrays

NumPy offers two useful methods for building up an array from smaller ones. Each takes a single argument, a list, or tuple whose entries are the arrays to be combined:

- **np.hstack** (horizontal stack): The resulting array has the same number of rows as the original arrays. The arrays to be stacked must have the same number of rows.
- **np.vstack** (vertical stack): The resulting array has the same number of columns as the original arrays. The arrays to be stacked must have the same number of columns.

Try these examples, and describe the results:

```
a = np.zeros( (2, 3) )
b = np.ones( (2, 3) )
h = np.hstack( [a, b] )
v = np.vstack( [a, b] )
```

Look at the contents of h and v, and compare their shapes with those of a and b.

2.2.5 Accessing array elements

Once you have created an array, you can access each of its entries individually. Try the following code at the command prompt:

```
A = np.array( [2, 4, 5] )
A[0]
A[1] = 100
print(A)
```

Array indices (offsets) are enclosed in square brackets, not parentheses.

Using round (ordinary) parentheses when requesting array elements is a common error for Python beginners.

The third line above changes just one entry in the array, but it may not be the entry you expect. In Python, the indices of lists, tuples, arrays, and strings all start with 0. Thus, `A[1]` is the *second* element of `A`. The same indexing scheme can access individual characters in strings. This is not the convention that math texts use for vectors and matrices.[6]

Indices in Python start at 0. If A is a one-dimensional array with N elements, the first element is `A[0]` *and the Nth element is* `A[N-1]`*. Asking for element* `A[N]` *will result in an error.*

A math text would refer to `A[N-1]` as A_N. The notation A_{-1} would probably not appear in most math texts, but Python interprets a negative index as an offset from the *end* of a list or array. More precisely, `A[-1]` will access the last entry of an array, and `A[-n]` refers to the nth element from the end of an array.

To understand how to access elements in a two-dimensional array, try the following:

```
A = np.array( [ [2, 3, 5], [7, 11, 13] ] )
A[0]
A[0][1]
A[1][2] = 999
```

In the second line above, `A[0]` returns an array whose elements are `[2,3,5]` (the first row). In the third line, `A[0][1]` then asks for the *second* item in *that* array, which is the number 3. NumPy arrays also understand an abbreviated indexing scheme:

```
A = np.array( [ [2, 3, 5], [7, 11, 13] ] )
A[0, 1]
A[1, 2] = 999
```

Both `A[i][k]` and `A[i,k]` return the entry at the intersection of row `i+1` and column `k+1`. A mathematics text would call this entry $A_{i+1,k+1}$. Again, Python's behavior is easier to understand if you think in terms of offsets: `A[i,k]` is the entry `i` steps down and `k` steps to the right from the upper-left corner of the array.

---

[6] T2 The convention in Python is inherited from an older programming language called C. In that language, the name of an array is linked to the location in memory where its first value is stored. An index is interpreted as an offset. Thus, to get the *first* entry in the array, we need an offset of *zero*: `A[0]`.

2.2.6 Arrays and assignments

An assignment statement for a variable, such as `f=2.5`, looks similar to an assignment statement for an element of an array, such as `A[1]=2.5`. However, these statements are handled differently in Python, and it is important to understand the distinction.

When we access and assign elements of an array, we are using methods of an `ndarray` object to view and modify its data.[7] In contrast, when we assign a value to a variable, as in `f=2.5`, Python binds the variable name to a new object. There is little difference in behavior until we have two names bound to the same object.

Suppose that we create two variables that point to the same `float` object and two variables that point to the same `ndarray` object:

```
f = 2.5
g = f
A = np.zeros(3)
B = A
```

Inspect all four variables, `A`, `B`, `f`, and `g`, before and after these commands:

```
g = 3.5
A[0] = 1
B[1] = 3
```

There is no effect on `f`, even though the value of `g` is changed. However, `A` and `B` remain equal to each other. The reason is that `A` and `B` are bound to the same `ndarray` object, and each used a method of that object to modify its data.

> *If multiple variables are bound to the same array, they can* ***all*** *use its methods to modify its data.*

This includes assignment of individual elements.

Lists behave in a similar way. Tuples are immutable objects that do not allow reassignment of their elements, so `A[0]=1` results in an error if `A` is a tuple.

There is another important difference between variable assignment and array assignment. If you type `A=1`, Python does not raise an error, even if `A` was previously undefined. However, if you try `A[0]=1` before defining `A`, Python returns an error message. The reason is that `A[0]=1` is attempting to use a method of an object named `A` to modify its data. Since no such object exists, Python cannot proceed. If you need to set the entries of an array one by one, you should first create an array of the appropriate size, for example, `A=np.zeros(100)`.

2.2.7 Slicing

Often you will wish to extract more than one element from an array. For example, you may wish to examine the first ten elements of an array or the 17th column of a matrix. These operations are possible using a technique called **slicing**.

In Python, the colon indicates slicing when it is part of an index. The most general command for slicing a one-dimensional array is `a[start:end:stride]`. This returns a new array whose entries are

---

[7] T2 The expression `A[1]` is shorthand for a method called `A.__getitem__(1)`, and `A[1]=2.5` calls the method `A.__setitem__(1,2.5)`.

```
a[start], a[start + stride], a[start + 2*stride], ..., a[start + M*stride]
```

where `M` is the largest integer for which `start + M*stride < end`.

> *The syntax for slicing is* `a[start:end:stride]`*. The stride comes last. If it is omitted, the default is* `1`*. If* `start` *or* `end` *are omitted, their defaults correspond to the first and last entries, respectively.*

If the stride is omitted, the second colon may also be omitted.

Thus, if you want to see the first ten entries in `a`, you could use the expression `a[0:10:1]`, or more concisely `a[:10]`. The same syntax can be used to slice along each dimension of a multidimensional array. For example, to get a slice of the third column of a matrix, you could use `A[start:end:stride,2]`.

Suppose that you have been given experimental data in an array `A` with two columns. For your analysis, you need to split the two columns of data into separate arrays. If you don't know the number of rows in `A`, you could start by finding it, and then assign the slices to separate variables:

```
N = np.size(A, 0)
x = A[0:N:1, 0]
y = A[0:N:1, 1]
```

A colon all by itself represents every allowed value of an index (that is, `start=0`, `end=-1`, `stride=1`). We can use this shortcut to specify our slices more concisely:

```
x = A[:, 0]
y = A[:, 1]
```

Your Turn 2D

Try the following commands to familiarize yourself with slicing:

```
a = np.arange(20)
a[:]
a[::]
a[5:15]
a[5:15:3]
a[5::]
a[:5:]
a[::5]
```

Explain the output you get from each line. Can you construct a slicing operation that returns only the odd entries of `a`?

Negative index values can also be used in slicing. This can be especially useful if you only want to see the last few entries in an array whose size is unknown. For example, `a[-10:]` will return the last ten elements stored in `a` (if `a` has at least ten elements).

An array slice can also appear in an assignment statement:

```
A = np.zeros(10)
A[0:3] = np.ones(3)
```

This code replaces a block of values in `A`, leaving the rest unchanged. (See Section 2.2.6.)

2.2.8 Flattening an array

An array may have any number of dimensions. In some cases, you may wish to repackage all of its values as a one-dimensional array. The function **`np.ravel`** does exactly this. It can be accessed as a NumPy function or an array method. NumPy arrays can also *flatten* themselves. Try the following commands, and inspect the results:

```
a = np.array( [ [1, 2], [2, 1] ] )
b = np.ravel(a)
c = a.ravel()
d = a.flatten()
```

Notice, in particular, that neither **`ravel`** nor **`flatten`** changes `a`. Each returns a one-dimensional array that contains the same elements as `a`. There is a significant difference between the methods, however. The **`flatten`** method returns a new, independent array; **`ravel`** returns an `ndarray` object with access to the *same* data as `a`, but with a different shape. To see this, compare the effects of `d[1]=11` and `b[2]=22` on all four arrays.

2.2.9 Reshaping an array

Flattening an array is just one way to change its shape. An array can be recast into any shape consistent with its number of elements by using **`np.reshape`** or an array's own **`reshape`** method. Try the following, and inspect each array:

```
a = np.arange(12)
b = np.reshape(a, (3, 4) )
c = b.reshape( (2, 6) )
d = c.reshape( (2, 3, 2) )
```

The **`reshape`** method takes a tuple of numbers, like `(3,4)` or `(2,3,2)`, as its argument. As long as the product of these numbers is equal to the number of elements in the original array, the command will return a new array. Like **`ravel`**, these **`reshape`** methods return an `ndarray` object with access to the same data as `a`, but with a different shape. Modifying the elements of any of the arrays in this example will affect all four of them.[8]

2.2.10 T2 Lists and arrays as indices

Python has an even more flexible method for slicing NumPy arrays: You can use a list where an index is expected. Using a list of integers will return an array that contains just the elements of the original array at the offsets specified by the list. Try

```
a = np.arange(10, 21)
b = [2, 4, 5]
a[b]
```

This method can be extremely useful if the list `b` was itself generated automatically. For example, `b` could be a list of time points at which we wish to sample a signal contained in `a`.

[8] T2 Array slicing, reshaping, and NumPy's **`ravel`** method each return a **view** of the original array. That is, the resulting object has access to the same data as the original array. Any changes to a view change the original object—and all other views of that object. See Appendix E.

You can also use a **Boolean** array (an array whose entries are either `True` or `False`) to select entries from another array of the same shape. This technique is called **logical indexing**. Try the following example:

```
a = np.arange(-10, 11)
less_than_five = (abs(a) < 5)
b = a[less_than_five]
```

The comparison in the second line returns an array with the same shape as `a` whose entries are `True` or `False`, depending on whether the particular element in `a` satisfies the comparison.[9] When `less_than_five` is used as an index to `a`, Python returns an array containing only those elements of `a` for which the corresponding element in `less_than_five` is `True`.

It is not necessary to create a named array to use as an index. The following line is equivalent to the last two lines of the example above:

```
b = a[abs(a) < 5]
```

2.3 STRINGS

Python can manipulate text as well as numbers. Next to an array, the second most important data structure for our purposes is the **string**. A string may contain any number of characters. You can create one with

```
s = 'Hello, world!'
```

Notice how `s` looks in the Variable Explorer.

The expression on the right side of the equals sign above is a **string literal**. The equals sign assigns this short sentence (without the quotes) as the value of `s`. Python allows you to use either a single quote (`'`) or a double quote (`"`) to define a string; however, you must begin and end the string with the same character.

A string starts and ends with a single quote or a double quote.

A single quote is the same as the apostrophe on your keyboard. Elsewhere on your keyboard there's a different key, the grave accent. You may be tempted to use it, as in `` `string' ``. Python won't understand that.

If you need an apostrophe inside a string, you can enclose the whole string in double quotes: `"Let's go!"` If you need both an apostrophe *and* double quotes, you will have to use a backslash (\) before the symbol that encloses the string. Write either `"I said, \"Let's go!\""` or `'I said, "Let\'s go!"'` More generally, inside a string, the backslash is an **escape character** that means, "Interpret the next character literally." In these examples, it instructs Python that the double quotes and the apostrophe, respectively, are part of a string, not the beginning or end of a string.[10]

A string may contain a collection of digits that looks like a number, for example, `s='123'`. However, Python *still considers such a value to be a string*, not the number 123. Try typing

---

[9] This is an example of *vectorized* computation, which will be discussed in Section 3.2.1.

[10] Another use of backslash is to code special characters, such as newline (`\n`) and tab (`\t`). You can even code backslash itself via `\\`. See page 54.

```
a = '123'
b = a + 1
```

Python issued a **TypeError** when you tried to add a number to a string. However, you can *convert* a string to a number. (This may be necessary if you are getting input from a keyboard.) Python can do sensible conversions like creating an integer from `"123"` or a floating point number from `"3.14159"`, but it does not know how to directly create an integer from a string that contains a decimal point. Try the following:

```
s = '123'
pie = '3.142'
x = int(s) + 1
y = float(pie) + 1
z_bad = int(s) + int(pie)
z_good = int(s) + int(float(pie))
```

Python does know how to "add" two strings together. Try this:

```
"2" + "2"
```

The result is not what you learned in kindergarten. Python uses the plus sign (+) to join (*concatenate*) strings:

```
s = 'Hello, world!'
t = 'I am Python.'
s + t
```

Your Turn 2E

> The result of the last evaluation doesn't look quite right. Replace the last line with `s+' '+t` and explain the operations that lead to the output.

Some Python functions require arguments that are strings. For example, graph titles and legends must be specified as strings. (See Section 4.3.) You may wish to include the value of a variable in such a string. If the value is not already a string, one option is to convert it. Just as Python can convert strings to integers and floating point numbers, it can convert numbers to strings. The built-in function **str** will try to create a string using whatever input you provide:

```
s = "Poisson distribution for $\\mu$ = " + str(mu_val)
```

This string can now be used by PyPlot to generate a graph title. In this expression,

- The + sign joins two strings.
- The first string is a literal. It contains some special characters that PyPlot can use to produce the Greek letter μ.[11]
- The second string is obtained from the current value of a variable named `mu_val`.

Python will display up to 16 digits when printing floating point numbers, and this is what the **str** command will generate in the example above if μ has that many digits. This may be satisfactory,

[11] T2 Python will interpret the text between dollar signs (`$...$`) as LaTeX typesetting instructions. The first backslash escapes the second one and prevents it from being interpreted as an escape character itself.

but Python offers much more control over the appearance of strings. You can control how numbers appear in strings by using special commands involving the percent sign (`%`) or the **`format`** method that every string possesses. The percent sign works in all versions of Python and is widely used, so we will briefly describe it below. However, the **`format`** method behaves more like other functions in Python, so we will discuss it first and use it throughout the remainder of the tutorial.

2.3.1 Formatting strings with the `format()` method

Let's look at some examples that use the **`format()`** method:[12]

```
# string_format.py
"The value of pi is approximately " + str(np.pi)
"The value of {} is approximately {:.5f}".format('pi', np.pi)
s = "{1:d} plus {0:d} is {2:d}"
s.format(2, 4, 2 + 4)
"Every {2} has its {3}.".format('dog', 'day', 'rose', 'thorn')
"The third element of the list is {0[2]:g}.".format(np.arange(10))
```

When evaluating a string's **`format`** method, Python interprets a pair of curly brackets (`{}`) as a placeholder, where values will be inserted. The arguments of **`format`** are expressions whose values will be inserted at the designated points. They can be strings, numbers, lists, or more complicated expressions. These examples illustrate several properties of the **`format`** method:

- An empty set of curly brackets will insert the next item in the series of expressions, as illustrated in the third line. The item can be formatted if the brackets contain a colon (`:`) followed by a formatting command. For example, "`{:.5f}`" means, "Format the current argument as a floating point number with 5 digits after the decimal point."
- Items can also be explicitly referenced by their location in the series of expressions passed to **`format`**. For example, the fourth line in the examples above defines `s` as a string with three placeholders to be filled later. It will take the second, then the first, then the third arguments and insert them at the designated locations. The syntax "`{1:d}`" means, "Insert the second argument here, and display it as a decimal (base 10) integer." (Python can also represent integers as binary, octal, or hexadecimal by using `{:b}`, `{:o}`, and `{:x}`, respectively.)
- Line 6 shows that not all arguments need to be used.
- The last line demonstrates how you can reference an individual item of an array passed to **`format`**: The *replacement field* `0[2]` refers to the third element of the first argument, and the *format specifier* `:g` then instructs Python to display a number using the fewest characters possible (general format). It may choose exponential notation, and it will leave off trailing zeros in floating point numbers.

Python's **`help(str)`** information provides basic descriptions of the available string methods. There are also many references online if you wish to explore string processing beyond the rudimentary introduction provided here. However, it can be quicker and more fun to

Experiment in the console to discover how things work.

---

[12] The hash symbol # in the following code snippet introduces a comment. (See Section 3.3.4.) The comment here tells you the name of the code snippet, so that you can find it in the book's online resources.

2.3.2 T2 Formatting strings with %

Try the following at the command prompt to see how the % syntax works:

```
# string_percent.py
"The value of pi is approximately " + str(np.pi)
"The value of %s is approximately %.5f" % ('pi', np.pi)
s = "%d plus %d is %d"
s % (2, 4, 2 + 4)
```

When a string literal or variable is followed by the % operator, Python expects that you are going to provide values to be formatted and inserted into that string. The desired insertion point(s) are indicated by additional % characters within the string. These can be followed by information about how each value should be formatted, as described in Section 2.3.1. As these examples show, you can insert multiple values into a string, and each value can be the result of evaluating an expression. You must provide the values in the order they are to appear in the string.

Thus, in the third line of the example above, "`%s`" means, "Insert a string here." Similarly, "`%.5f`" means, "Insert a floating point number here with 5 digits after the decimal point." In the third line of the example, "`%d`" means, "Insert a decimal (base 10) integer here."

In addition to formatting strings, the percent sign can also function as an arithmetic operator in Python. In an expression like `5%2`, the percent sign represents the **modulo operation**. It returns the remainder upon division. (Try it.) Although this particular operation is not used frequently in physical modeling, we mention it because it can cause puzzling behavior when you have a bug that involves a percent sign.

This chapter has introduced the most common Python data structures in scientific computing. This is enough to get started with array processing and string manipulation. However, to get the most out of Python we need to do more than enter instructions at the command line. It is time to look at the flow of operations in a chain of instructions—that is, at computer *programs*.

CHAPTER 3

Structure and Control

When you come to a fork in the road, take it.
— Yogi Berra

The previous chapter introduced some useful structures for storing and organizing data. To utilize them effectively, we now need to automate repetitive operations on the data. This chapter describes how to group code into repeated blocks (looping) and contingent blocks (branching), and how to assemble code blocks into reusable computer programs called scripts.

3.1 LOOPS

So far, we have described Python as a glorified calculator with some string processing capabilities. However, we can continue to build on what we have learned and do increasingly complex tasks. In this section, we will explore two **control structures** that will allow us to repeat a set of operations as many times as we need: **`for`** loops and **`while`** loops.

3.1.1 `for` loops

Instead of solving a quadratic equation, let's go a step further and create a *table* of solutions for various values of `a`, holding `b` and `c` fixed. To do this, we will use a **loop**. Try typing the following in the IPython console:

```
# for_loop.py
b, c = 2, -1
for a in np.arange(-1, 2, 0.3):
    x = (-b + np.sqrt(b**2 - 4*a*c)) / (2*a)
    print("a= {:.4f}, x= {:.4f}".format(a, x))
```

(Hit `<Return/Enter>` on a blank line to terminate the loop and run it.)

Note the colon after the **`for`** statement. This is essential. It tells Python that the next lines of indented code are to be associated with the **`for`** loop. The colon is also used in **`while`** loops and **`if`**, **`elif`**, and **`else`** statements, which will be discussed shortly.

The keyword **`for`** instructs Python to perform a block of code repeatedly. It works as follows:

1. The function `np.arange(-1,2,0.3)` indicates a series of values with which to execute the indented block of code. Python starts with the first entry and cycles over each value in the array.
2. Python initially assigns `a` the value -1. Then, it evaluates a solution to the quadratic equation and prints the result. (Without the **`print`** statement, nothing would be displayed as the calculation proceeds.)

3. The end of the indented code block tells Python to move back to the beginning of the loop, update the value of a, and execute the block *again.* Eventually, Python reaches the end of the array. Python then jumps to the next block of unindented code (in this case, nothing). We say that it "exits the loop."

Note that a is an ordinary variable whose value can be accessed in calculations. It is generally not good practice to modify its value within the loop, however. When Python reaches the end of an indented block, it will continue on to the next value in the array and discard any changes you made to the value of a.

The example above illustrates a very important feature of Python:

Blocks of code are defined ***only*** *by their indentation.*

In many programming languages, special characters or commands separate blocks of code. In Java, C, or C++, the instructions in a **for** loop are enclosed in curly brackets ({...}). In other languages, the end of a block of code is designated by a keyword (for example, end in MATLAB). In Python, the indentation of a statement—and nothing else—determines whether it will be executed inside or outside a loop. This forces you to organize the text of your programs exactly as you intend the computer to execute it. You can tell at a glance which commands are inside a loop and where a block of code ends, and your code is not cluttered with dangling curly brackets or a pyramid of end statements at the end of a complex loop.

How much should you indent? The only requirements in Python are the following:

(*i*) Indentation consists of blank spaces or tabs.

(*ii*) Indentation must be consistent within a block. (For example, you cannot use four spaces on one line and a single tab on the next, even if the tab appears to indent the block by four spaces.)

(*iii*) The indentation level must increase when starting a new block, and go back to the previous level when that block ends.

Try this modified set of instructions and explain why the output differs from before (before typing the last line, hit backspace to undo the automatic indentation):

```
b, c = 2, -1
for a in np.arange(-1, 2, 0.3):
    x = (-b + np.sqrt(b**2 - 4*a*c)) / (2*a)
print("a= {:.4f}, x= {:.4f}".format(a, x))
```

We can tell by looking that this version of the loop will cycle over the values of a, and *then* display the final values of a and x once Python exits the loop.

IPython's command line interpreter aids you in setting up indentation. When you typed the first line of the **for** loop and hit <Return/Enter>, the next line was automatically indented. When you write your own programs, you will have to make sure statements are indented properly. (Many semi-intelligent editing programs, including the Editor in Spyder, will also help you with indentation.)

If the body of a loop is very brief, you can just write it on the same line as the **for** statement after the colon, and forget about indentation:

```
for i in range(1, 21): print(i, i**3)
```

3.1.2 while loops

A second example of a control structure is useful when you wish to repeat a block of code as long as some general condition holds but you do not know how many iterations will be required. A **while** loop will exit the first time its condition is **False**. For example, you could calculate solutions to a quadratic equation until the discriminant changes sign as follows:

```
# while_loop.py
a, b, c = 2, 2, -1
while (b**2 - 4*a*c >= 0):
    x = (-b + np.sqrt(b**2 - 4*a*c)) / (2*a)
    print("a = {:.4f}, x = {:.4f}".format(a, x))
    a = a - 0.3
print("done!")
```

Note that we now need to change the value of a explicitly within the loop (line 6), because we are not iterating over an array. Also, be aware that a different choice of values for a, b, and c could result in an infinite loop. (See Section 3.1.4 below.)

As with a **for** loop, the indentation of statements tells Python which commands to execute inside a **while** loop; it also tells Python at what point to pick up again after the loop has completed. In the example above, the code prints out a short message after finishing the loop.

3.1.3 Very long loops

Some calculations take a long time to complete. While you're waiting, you may be wondering whether your code is really working. It's a good idea to make your code provide updates on its progress. If you have a loop like `for ii in range(10**6):`, then inside the loop you may wish to place

```
if ii % 10**5 == 0: print("{:.0f} percent complete".format( 100*ii/10**6 ))
```

Here the percent sign is being used to determine a remainder.

You may want to do some other task while waiting for your code to complete. In that case, it's convenient to get audible bulletins. The simplest is `print("\a")`. Try it!

3.1.4 Infinite loops

When working with loops, there is the danger of entering an **infinite loop** that will never terminate. This is usually a bug, and it is useful to know how to halt a program with an infinite loop without quitting Spyder itself.

The easiest way to halt a Python program is to issue a **KeyboardInterrupt** by typing <Ctrl-C>. This will work on most programs and commands. Try it now. At the IPython command prompt, type

```
while True: print("Here we go again ...")
```

When you hit <Return/Enter>, Python will enter an infinite loop. Halt this loop by typing <Ctrl-C>.

<Ctrl-C> can also be used to halt commands or scripts that are taking too long, or lengthy calculations that you realize are using the wrong input. Just click in the IPython console pane, and type <Ctrl-C>. There is also a red STOP■ button in the upper-right corner of the IPython console that will issue a **KeyboardInterrupt** from within Spyder (see Figure 1.1, page 6). Run the infinite loop command again, and halt it with the STOP■ button.

It may take a while for Python to respond, depending on what it is doing, but `<Ctrl-C>` or the STOP■ button will usually halt a program or command. However, if Python is completely unresponsive, you may have to exit Spyder or force it to quit with the `Restart kernel` option in the OPTIONS⚙ menu to the right of the STOP■ button on the IPython console pane (again see Figure 1.1, page 6). If this happens, you could lose everything you have not saved, so be sure to save your work frequently.

3.2 ARRAY OPERATIONS

One reason to use arrays is that NumPy has a very concise syntax for handling repetitive operations on arrays. NumPy can directly calculate the square root of every element in an array or the sum of each column in a matrix much more rapidly than a **`for`** loop designed to do the same thing. Applying an operation to an entire array instead a single number (scalar) is called **vectorization**. An operation that combines elements of an array to create a smaller array—like summing the columns of a matrix—is called **array reduction**. Let's look at examples of both of these operations.

3.2.1 Vectorizing math

We can calculate the solutions to the quadratic equation in Section 3.1.1 by using the following code:

```
# vectorize.py
b, c = 2, -1
a = np.arange(-1, 2, 0.3)
(-b + np.sqrt(b**2 - 4*a*c)) / (2*a)
```

We type the last command exactly as we would for a single value of `a`, but Python evaluates the operation for every element of `a` and returns the results in an array.

To evaluate the expression on the fourth line, Python essentially carries out the following steps:[1]

1. Python starts with the innermost subexpression, first calculating `b**2` and saving the result. In evaluating `4*a*c`, it notices that the array `a` is to be multiplied by 4. Python interprets this as a request to multiply *each element* by 4. This is the standard interpretation in math: When you multiply a vector by a scalar, each component is multiplied by that quantity. Similarly, Python then multiplies the resulting array by `c`.
2. After evaluating `4*a*c`, Python sees that you've asked it to combine this result with the single quantity `b**2`. *Unlike* standard math usage (where it does not make sense to add a vector and a scalar together), Python interprets this as a request to create a new array, in which the entry indexed by k is equal to `b**2 - 4*a[k]*c`. That is, addition and subtraction of an array and a scalar are also evaluated item by item.
3. Python then passes the array generated in step **2** to the **`np.sqrt`** function as an argument. NumPy's square root function will accept a single number, but like many functions in the NumPy module (including **`sin`**, **`cos`**, and **`exp`**), it can also act on an array. The function acts item by item on each entry and returns a new array containing the results.[2]

---

[1] Python may actually use a more efficient procedure.

[2] The module we are using is important: The `sqrt` function in the **`math`** module will not accept an array as an argument.

4. Python adds `-b` to the array from **3** using the same rule as in **2**.
5. Python divides the array from **4** by `2*a`. This operation involves two arrays and will be discussed in a moment. It is also evaluated item by item.

The code in `vectorize.py` is a *vectorized* form of the earlier **for** loop. Vectorized code often runs much faster than equivalent code written with explicit **for** loops, because clever programmers have optimized NumPy functions to run "behind the scenes" with compiled programs written in C and FORTRAN, avoiding the usual overhead of Python's interpreter.

Vectorization can also help you write clearer code. Long, rambling code can be hard to read, making it difficult to spot bugs. Of course, extremely dense code is also hard to read. Your coding style will evolve as you get more experienced. Certainly if a code runs fast enough with **for** loops, then there's no need to go back and vectorize it.

Not every mathematical operation on an array will act item by item, but most common operations do. Suppose that you wanted to graph the function $y = x^2$. You could set up an array of x values by `x=np.arange(21)`. Then, `y=x**2` or `y=x*x` gives you what you want.

We can now finish explaining the example at the start of this section. The variable `a` is an array; therefore, `-b + np.sqrt(b**2-4*a*c)` and `2*a` are arrays, too. When two NumPy arrays of the same shape are joined by `+`, `-`, `*`, `/`, or `**`, Python performs the operation on each pair of corresponding elements, and returns the results in a new array of the same shape.

Your Turn 3A

a. We often wish to evaluate the function $y = \mathrm{e}^{-(x^2)}$ over a range of x values. Figure out how to code this in vectorized notation.
b. We often wish to evaluate the function $\mathrm{e}^{-\mu}\mu^n/(n!)$ over the integer values $n = 0, 1, \ldots, N$. Here, the exclamation point denotes the factorial function. Figure out how to code this in vectorized notation for $N = 10$ and $\mu = 2$. (You may want to import the **factorial** function from SciPy's collection of special functions, **scipy.special**.)

The item-by-item operations work equally well with multidimensional arrays. Again, the arrays to be combined must have the same shape. The expressions `a+b` and `a*b` generate error messages unless `a.shape==b.shape` is **True** or `a` and `b` can be "broadcast" to a common shape. (For example, you can always add a number to an array of any shape. For more information, try a web search for `numpy broadcasting`.)

Most math operations on NumPy arrays are performed item by item.

Vectorized operations apply only to NumPy arrays. Most mathematical operators are not defined for Python lists, tuples, or strings. Some are, but they do not carry out arithmetic. Execute the following commands to see why we do not use Python's `list` data structure:

```
x = [1, 2, 3, 4, 5, 6]
2 * x
x + 2
x * x
x - x
```

If you "add" two lists, two tuples, or two strings with a +, the result will be to join, or concatenate, them. Python's `list` objects are useful in many computing applications, but fast mathematical calculation on large collections of numbers is not one of them.

3.2.2 Matrix math

Sometimes, instead of item-by-item operations, you may want to combine arrays by the rules of matrix math. For example, the "dot product" of two vectors (or, more generally, matrix multiplication) requires a special function call. Compare the output of the two operations below:

```
a = np.array( [1, 2, 3] )
b = np.array( [1, 0.1, 0.01] )
a*b
np.dot(a, b)
```

The number of elements in `a` equals the number of elements in `b`, so the dot product is defined. In this case, it is the single number $\boldsymbol{a} \cdot \boldsymbol{b} = a_1 b_1 + a_2 b_2 + a_3 b_3$.

3.2.3 Reducing an array

In contrast to ordinary functions like **`sin`**, which act item by item when given an array, some functions combine elements of an array to produce a result that has fewer elements than the original array. Sometimes the result is a single number. Such operations are said to **reduce** an array.

A common array reduction is finding the sum of the elements in each row or column, or the sum of all the elements in the array. Try

```
a = np.vstack( (np.arange(20), np.arange(100, 120)) )
b = np.sum(a, 0)
c = np.sum(a, 1)
d = np.sum(a)
```

The function `np.sum(a,n)` takes an array `a` and an integer `n` and creates a new array. Each entry in the new array is the sum of the entries in `a` over all allowed values of index number n, holding the other indices fixed. In the example above, `n=0` specifies the first axis; thus, `b` contains the column sums of `a`. Setting `n=1` specifies the second axis; thus `c` contains the row sums of `a`. When no index is given, as for `d`, **`np.sum`** adds up *all* of the elements in the array.

Python also has a built-in **`sum`** function that does not work the same way. Try the example above using **`sum`** without the "**`np.`**" prefix to see what happens, then look at `help(sum)` and `help(np.sum)` to understand the difference between the two functions.

You can use experimentation and the **`help`** command to explore the useful related functions **`np.prod`**, **`np.mean`**, **`np.std`**, **`np.min`**, and **`np.max`**. Alternatively, every array comes with methods that evaluate these functions on the array's own data. For instance, the example above could also have been written as follows:

```
a = np.vstack( (np.arange(20), np.arange(100, 120)) )
b = a.sum(0)
c = a.sum(1)
d = a.sum()
```

See if you can explain how the following code calculates 10!:

```
ten_factorial = np.arange(1, 11).prod()
```

3.3 SCRIPTS

You can do many useful things from the IPython Console, which is what we have been using up to this point. However, retyping the same lines of code at the command prompt over and over is tedious, and many tasks involve code that is far more complex than the examples above. Your code will go through many versions as you get it right, and it's tiresome to keep retyping things and making new errors. You'll want to work on it, take a break, return to it, close Spyder, move to a different computer, and so on. And you will wish to share code with other people in a "pure" form, free from all the typos, missteps, and output made along the way. For all these reasons, you will want to do most of your coding in the form of scripts.

A **script** is simply a text file that contains a series of commands. You edit a script in a text editor, and then execute it by calling it with a single command or a mouse click.

3.3.1 The Editor

You can use Spyder's Editor to create and edit scripts. To use the Editor, simply click on its pane or use the keyboard shortcut <Cmd-Shift-E>. If the Editor is not open, the keyboard shortcut will open it. To create a script, you can open a new document from the File menu, you can click on the blank page icon at the upper left, or you can use the keyboard shortcut <Cmd-N>.[3]

Write a script that contains any of the code fragments above. As you enter text in a script, hitting <Return/Enter> only moves the cursor to the next line. No commands are executed because Python has no knowledge of your script yet. It is just a text file somewhere in your computer's memory.

When you finish editing, you can execute your file by clicking the RUN▶ button, selecting "Run" from the Run menu at the top of the window, or by using the shortcut key <F5>. Python executes the code in the Editor's active tab and displays any output in the IPython console. Before it does that, however, it saves the code, prompting you for a file name if necessary. It is customary to end Python code file names with the extension .py.[4] You can also execute the code by typing **%run** followed by your file name (with or without the .py extension) at the command prompt, but you will need to import modules and define variables within the script instead of relying on what you may have already done in the IPython console.

> *It's good practice to type* **%reset** *at the command prompt before running any script.*

That way, you know exactly what state Python is in when the script begins. Keep in mind that **%reset** will delete all imported modules. (See Section 1.2.1.) Thus, you will probably want to include the following lines at the top of *every* script you write:

```
import numpy as np
import matplotlib.pyplot as plt
```

You can have Spyder add these lines to every new script by default. See Appendix A.

---

[3] On Windows systems, use <Ctrl-Shift-E> and <Ctrl-N>. You can also use IPython to run scripts that you create outside of Spyder. See Section 3.3.2.

[4] When naming your script, do not use a Python reserved word (for.py), nor the name of a module you may want (numpy.py).

3.3.2 [T2] Other editors

If you prefer another text editor, you can edit and save files outside of Spyder and then run them within Spyder. Give your code file the extension `.py` (not `.txt`).

You can use the **%run** command at the IPython command prompt to execute any plain text Python file in the current directory. Formatted text files such as `.rtf`, `.doc`, or `.docx` files won't work, however. It's best to use a "plain text" editor that has three key features:

- Your editor should have an *autoindent* feature that will ensure consistent spacing in indented blocks of code. Many text editors also allow you to convert tabs into an equivalent number of spaces. If you do not use either feature, you could accidentally mix tabs with spaces in indented blocks. The code may look correctly indented on the screen, but it won't run properly.
- Your editor should also have *syntax highlighting.* This useful feature displays Python commands and keywords in different colors or fonts, allowing you to spot some coding errors at a glance. If your editor supports syntax highlighting, it will operate automatically whenever you open a file with extension `.py`.
- Your editor should support *bracket matching,* which identifies unmatched brackets such as a "`(`" without a "`)`", or an extra "`)`".

Vim and Emacs are classic text editors included with most Unix and Linux systems. Atom is a more recent open-source editor for all operating systems (`atom.io`). Other options include Notepad++ for Windows (`notepad-plus-plus.org`), TextWrangler for macOS (`www.barebones.com`), and gedit for Linux (`wiki.gnome.org/Apps/Gedit`), all of which are free.

If you are running IPython *without* Spyder—for instance, from your operating system's command line—then IPython's magic commands can accommodate you in editing files and running scripts. You can open a script for editing by typing **%edit** `my_script.py`. This will open `my_script.py` in your default text editor. Once you have created or edited the file, save it and quit the editor. IPython may automatically run your file. You can also run the script by typing **%run** `my_script.py`.

Finally, you can execute a Python script directly from your operating system's command line, outside of any IDE, with the command

```
$ python my_script.py
```

3.3.3 First steps to debugging

If you make an error in your code while you are typing it in the Editor, Spyder may catch it before you even click [RUN ▶]. Lines with potential errors are flagged by yellow triangles and red diamonds with an exclamation point inside. A red diamond means your script will not run; a yellow triangle means it may not run.[5] Type the following lines in the Editor to see the two types of errors:

```
s = str 3
prnt(s)
```

(You may need to save the file before Spyder checks it for errors.) Spyder recognizes that you are using the known function **str** incorrectly in the first line (a *syntax error*) and flags it with a red diamond. Move the mouse cursor over the red diamond to get a brief explanation. The information provided is sometimes cryptic, but just knowing the general location of an error is quite helpful.

---

[5] [T2] Spyder's syntax checking is performed by the `pyflakes` module, which can be run independently of Spyder. Another popular syntax checker is `pylint`.

Fix the first line so that s is assigned the string representation of 3. Now, a yellow triangle will appear next to the second line. (If there is a fatal error, Spyder will not bother warning you about other potential trouble. Thus, if there is more than one fatal error, Spyder will usually recognize only the first.) In this case, Spyder is not aware of any function called `prnt`, but you may have defined it somewhere else. If you click [RUN▶], Spyder will attempt to execute the code. When Python encounters the `prnt` command, it discovers that no such name is defined, and so it *raises an exception* called **`NameError`**. That is, it prints a *runtime error message* in the IPython console and aborts any further processing.

> *A lot of information may print out when Python encounters an error. Generally, the last few lines are the most helpful.*

The last portion of the error message will cite the line number where Python first noticed something wrong, display a portion of the script near this line, and describe the type of error encountered. The script may contain other errors, but Python quits when it hits the first one. Fix it and try again.

> *The most common cause of* **`NameError`** *is misspelling.*

Variable names are case sensitive, so inconsistent case is a form of spelling mistake. For example, if you define `myVelocity=1` and later attempt to use `myvelocity`, Python treats the second instance as a totally new, undefined variable. Scanning the Variable Explorer can help you find such errors.

> *Another common cause of* **`NameError`** *is forgetting to prefix a function name with the name or nickname of the module containing it.*

Runtime errors generate messages in the IPython console when you run the code. Most cause Python to halt immediately. A few, however, are "nonfatal." For example, try

```
for x in np.arange(-1, 8):
    print(x, np.log(x))
```

Python finds nothing syntactically incorrect here, but NumPy recognizes a problem when it attempts to calculate the first two values. It gives these values names in this case (**`nan`** for "not a number" and **`-inf`** for "infinity"), but it knows these values could cause problems later, so it issues two **`RuntimeWarning`** messages, and even explains what happened.[6]

Still other situations generate no message at all—you just get puzzling results. You will encounter many situations of this sort, and then you'll need to *debug* your code.

Whole books are written about the art of debugging. When a code generates messages or output that you don't understand, you need to look for clues. Here are some ideas:

- Read your code carefully. This is often the quickest route to insight.
- Build your code slowly. Most scripts are designed to execute a complicated task by breaking it into a series of simple tasks. Make sure each step does exactly what it is supposed to. If every step does exactly what it is supposed to do and your code still does not work, you have a *theory bug,* not a *programming bug.*
- Start with an easier case. You're using a computer because you can't do the problem by hand, but maybe there's another case that you *can* do by hand. Adapt your code for that case (perhaps just a matter of changing a few parameter values), and compare the output to the answer you know to be correct.

[6] Not all modules are as forgiving. Try importing the **`math`** module and using its `log` function inside the loop instead.

- Probe your variables. After a code finishes (or terminates with an error), all of its variables retain their most recent values; check them to see if anything seems amiss.
- Insert diagnostics. Somewhere prior to where you suspect there's an error, you can add a few lines that cause Python to print out the values of some variables *at that moment* in the code's execution. When you run the program again, you can check whether they're what you expect.
- Be proactive as you write. Make sure that what you think *should be* true actually *is* true when a command is executed. Python offers a very useful tool for this: the **assert** statement. At any point in your code, you can insert a line of the form

```
assert (condition), "Message string"
```

  If the condition is **True**, then Python will continue executing your code. If it is **False**, however, Python will stop the code, give an **AssertionError**, then print your message. For example, in the code above, we could insert the following line above the **print** statement:

```
assert (x > 0), "I do not know how to take the log of {}!".format(x)
```

  When the code runs, it will not only stop when it tries to take the logarithm of a bad value—it will also print out the value that caused the problem. Of course, you can't foresee every possible exceptional case. But you will develop a sense of what bugs are likely after some practice.
- Explain the code line-by-line out loud to another person or inanimate object. The latter practice is often called *Rubber Duck Debugging.* Professional developers commonly force themselves to explain malfunctioning code to a rubber duck or similar totem to avoid having to involve another developer. The act of describing what the code is *supposed* to do while examining what it actually does quickly reveals discrepancies.
- Ask a more experienced friend. This may be embarrassing, because almost all coding errors appear "stupid" once you've found them. But it's not as embarrassing as asking your instructor. And either one is better than endlessly banging your head on the temple wall. You need to become self-reliant, eventually, but it doesn't happen all at once.
- Ask an online forum. An amazing, unexpected development in human civilization was the emergence of sites like `stackoverflow.com` where people pose queries at every level, and total strangers freely help them. The turnaround time can be rather slow; however, your question may already be asked, answered, archived, and available.
- Learn about *breakpoints.* This topic goes beyond the scope of this tutorial, but there may come a day when your code is complex enough to need it.

Maybe the single most important point to appreciate about debugging is that *it always takes longer than you expect.* The inevitable corollary is

Don't wait till the day before an assignment is due to start.

Some puzzles don't resolve until your subconscious has had time to unravel them. If you need help from a friend, lab partner, or instructor, that takes time too. Respect the subtlety of coding, and give yourself enough time.

3.3.4 Good practice: Commenting

Another key advantage of writing scripts is that you are free to include as many remarks to your reader—and yourself—as you like.

Let's apply our example of calculating solutions to the quadratic equation by giving it some physical meaning. Suppose you are planning a prank on a friend and need to know how long a snowball remains in the air when thrown upward with a particular initial speed. You recall from introductory physics that the height of a ball thrown upward is

$$y(t) = y_0 + v_0\, t - \frac{1}{2}\, g\, t^2.$$

After you recall the meaning of all the symbols, you realize that you need the value of t for which $y(t) = 0$. Rearranging things a bit, you obtain

$$\frac{1}{2} g t^2 - v_0 t - y_0 = 0.$$

This is the quadratic equation we already solved, but now the parameters all have physical meanings.

You can use the following code to plan your prank:

```
# projectile.py
# Jesse M. Kinder -- 2017
"""
Calculate how long an object is in the air when thrown from a specified height
with a range of initial speeds assuming constant acceleration due to gravity:
    0.5 * g * t**2 - v0 * t - y0 = 0
"""
import numpy as np
import matplotlib.pyplot as plt

#%% Initialize variables.
initial_speed = 0.0            # v0 = initial vertical speed of ball [m/s]
impact_time = 0.0              # t = time of impact [s] (computed in loop)

#%% Initialize parameters.
g = 9.8066                     # Gravitational acceleration [m/s^2]
initial_height = 2.0           # y0 = height ball is thrown from [m]
speed_increment = 5.0          # Speed increment for each iteration [m/s]
cutoff_time = 10.0             # Stop computing after impact time exceeds cutoff.

#%% Calculate and display impact time.  Increment initial speed each step.
#   Repeat until impact time exceeds cutoff.
while impact_time < cutoff_time:
    # Use quadratic equation to solve kinematic equation for impact time:
    impact_time = (np.sqrt(initial_speed**2 + 2 * g * initial_height) \
            + initial_speed) / g
    print("speed= {} m/s; time= {:.1f} s".format(initial_speed, impact_time))
    initial_speed += speed_increment
print("Calculation complete.")
```

(The syntax `x+=1` is a concise and descriptive shorthand for `x=x+1`. Other arithmetic operations can be called in a similar way. For example, `x*=2` will double the value of `x`.)

Compare this code to what was written in Section 3.1.1. The code here accomplishes a similar task, but the purpose and function of the script are now easy to understand because of the comments, whitespace, and meaningful variable names.

- The opening lines now tell us who wrote the program, when, and why.
- The first two commands import NumPy and PyPlot (which will be described later).
- Lines 11–19 assign values and meaningful names to the parameters and variables. (By "parameter" we simply mean a variable whose value is not going to change throughout the execution of the code.) Comments describe the role of each.
- The first parts of lines 12–13 and lines 16–19 are code, but everything after the hash signs (#) is commentary explaining the meaning of the assignments.
- Lines 21–22 introduce the main loop of the program and explain the condition for exiting the loop.

The script is certainly longer than the bare code introduced in Section 3.1.1, but which is easier to read? Which would *you* rather read, if you wanted to reuse this code after a month or two of not working on it?

> *Every coder eventually learns that good comments save time in the long run… as long as comments are updated to reflect changes in the code.*

You will probably ignore this advice until you return to a script you wrote several weeks earlier, cannot figure out what any of it means, and have to rewrite the whole thing from scratch. At least you were warned.

The script makes use of two different types of comments: One type is enclosed within triple quotes ("""), and the other is preceded by the hash sign (#). The hash sign is an *inline* comment character. Python will ignore everything that follows a hash sign on a single line. Comments can start at the beginning of a line or in the middle. Placing a short comment on the same line as the code being discussed is a good way to ensure that you'll remember to update the comment if you change the line later.

Some comments begin with the hash sign followed by two percent signs (#%%). These mean nothing special to Python, but they divide the code into logical units called **cells**. Besides looking good on the screen (try it), this structuring allows you to run individual cells separately within Spyder. To see how this works, click anywhere inside such a cell, then click the [Run cell] button (immediately to the right of the [Run▶] button) or use the shortcut <Ctrl-Enter>. Also try the alternative [Run cell and advance] (immediately to the right of the [Run cell] button), which has the shortcut <Shift-Enter>. Breaking code into cells by using comments is another useful debugging tool.

Triple quotes provide a convenient way to create comments or strings that span multiple lines. They can also be used to create special kinds of comments called documentation strings, or **docstrings** for short. A string enclosed in triple quotes becomes a docstring when it is placed at the beginning of a file or immediately after a function declaration.[7] Anywhere else in the file, a string enclosed in triple quotes is treated as an ordinary string.

Python uses docstrings to provide information about modules and functions you write. For example, the docstring in a function can tell a user what arguments are required, what the function will return, and so on. The script above is not a very useful module, but to see how the **help** feature works with the docstrings you write, you can import it anyway. Type `import projectile` (the file name without the `.py` extension) at the command prompt. Now type `help(projectile)`. You should see the text between the triple quotes.

Here is a final comment about comments: As you hack together a workable code, you may want to try out some temporary change, but preserve the option to revert your code to its preceding

[7] Section 6.1 discusses user-defined functions.

version. A simple method is simply to *keep* the old lines of code, but "comment them out" so that they become invisible to Python. If you want them back later, simply "uncomment" them. Spyder (like most other editors) offers a menu item to comment or uncomment a selected region of text, or you can use the shortcut `<Cmd-1>`.

The sample script above also illustrates proper use of indentation and other forms of *whitespace* (spaces, tabs, and blank lines).

- Proper indentation is crucial in Python. It also makes the code easier to read and understand. There is no ambiguity about which lines of code belong to the **`while`** loop and which do not.
- Python ignores blank lines, but they make the code much easier to read and interpret by dividing it into logical units.

3.3.5 Good practice: Using named parameters

Why did we clutter the code above with a variable name like `initial_height` instead of `y0`? Why did we bother to name the fixed parameters at all? We could have "hard coded" everything, replacing `g`, `initial_height`, and `speed_increment` by the numerical values `9.8066`, `2.0`, and `5.0` everywhere. There are at least four reasons:

1. We have separated what the code does from its input. This type of abstraction allows you to easily adapt your code for other purposes. You can run the same calculation with different values for the fixed parameters, embed the main loop of this code *inside* a loop in another program that evaluates *many* values of `initial_height`, or use the main loop inside a *function* that accepts the parameter values as input. Keeping them symbolic helps.
2. Using meaningful names for quantities improves your code's readability and clarity, even if those quantities stay fixed. Would you be as likely to understand the purpose of the program if there were no comments and we had just used short, uninformative variable names like `x1`, `x2`, and `x3`? You can probably recall taking notes in a physics or math class and momentarily forgetting what quantity a particular symbol stands for. It is much faster to write y_0 on the chalkboard or in your notebook, but the meaning is not always clear. There is no ambiguity about what the variable `initial_height` represents.
3. Naming parameters also keeps their values distinct, even when they are numerically equal. What if there were two parameters with different physical meanings that had the same numerical value of `5.0`? If you give variables and parameters different, meaningful names, then you won't get confused by such coincidences.
4. Using named parameters to control the *size* of a calculation is extremely useful. For instance, you may intend to set up ten arrays with five million entries each, and perform a thousand calculations on the collection of arrays. Such a code could take a long time to execute. It would waste a lot of time to wait for it to execute every time you fix a small bug or add a feature. It's better to set up some parameters (perhaps called `num_elements` and `max_iterations`) near the beginning of the script that control how big the calculation is going to be. You can choose small values for the development phase, and switch to the larger desired values only when you near the final version. Moreover, you can be certain that every array in the calculation will have the same number of elements.

These points are part of a bigger theme of coding practice:

Don't duplicate. Define once, and reuse often.

That is, if the same parameter appears twice, give it a name and set its value *once*. If you don't implement this principle, then when you wish to change a value, inevitably you will find and change all but one instance. The one you missed will cause problems, and it could be hard to find. Worse yet, you may accidentally change something else!

Later we will see how "don't duplicate" can be applied not just to parameter values but also to code itself when we discuss functions in Section 6.1.

3.3.6 Good practice: Units

Most physical quantities carry units, for example, 3 cm. Python doesn't know about units; all its values are pure numbers. If you are trying to code a problem involving a quantity L with dimensions of length, you'll need to represent it by a variable `length` whose value equals L *divided by some unit.* That's fine, as long as you are consistent everywhere about what unit to use.

You can make things easier on yourself if you include a block of comments at the head of your code declaring (to yourself and others) the variables you'll use, and the units chosen to make them pure numbers:

```
# Variables:
# -----------------------------------------------------
#    length          =    length of the microtubule [um]
#    velocity        =    velocity of motor [um/s]
#    rate_constant   =    rate constant [1/days]
...
```

(The notation `um` is an easy-to-type substitute for μm.) As you work on the code, you can refer back to this section of your comments to keep yourself consistent. Such fussy hygiene will save you a lot of confusion someday.

Later on, when you start to work with data files, you need to learn from whoever created the file what units are being used to express quantities. Ideally this will be spelled out in a text file (perhaps called `README.txt`) accompanying the data file, or in the opening lines of the data file. (You should document data files that you create in this way.)

3.4 CONTINGENT BEHAVIOR: BRANCHING

We have now explored several ways to automate repetitive calculations. Two of these methods, **`for`** loops and **`while`** loops, used control structures of the Python language to repeat a block of code. The **`while`** loop above modified its behavior on the fly when it decided whether to execute its block of code again depending on the result of an intermediate calculation. Every programming language, including Python, has a more general mechanism for specifying contingent behavior, called **branching**. Try entering this code in the Editor and running it:[8]

```
# branching.py
""" This script illustrates branching. """

import numpy as np, matplotlib.pyplot as plt
```

[8] In Python 2, **`input`** attempts to evaluate its argument as a Python expression. To run this script, users of Python 2 should replace **`input`** with `raw_input` or redefine **`input`** as `raw_input`. See Appendix D.

```
for trial in range(5):
    userInput = input('Pick a number: ')
    number = float(userInput)
    if number < 0:
        print('Square root is not real.')
    else:
        print('Square root of {} is {:.4f}.'.format(number, np.sqrt(number)))
    userAgain = input('Another [y/n]? ')
    if userAgain != 'y':
        break
#%% Check whether the loop terminated normally.
if trial == 4:
    print('Sorry, only 5 per customer.')
elif userAgain == 'n':
    print('Bye!')
else:
    print('Sorry, I did not understand that.')
```

3.4.1 The `if` statement

The code above illustrates several new ideas:

- The **`input`** function displays a prompt in the IPython console, then waits for you to type something followed by `<Return/Enter>`. Its return value is a string containing that input.[9] In this case, we assign that string to a variable called `userInput`, then convert it to a `float` for use later in the program.
- Line 9 contains an **if statement**. The keyword **`if`** is followed by a logical (*Boolean*) expression and a colon. If Python evaluates the expression and finds that it's **`True`**, then it executes the next indented block of code. If the expression is **`False`**, Python skips the block of code.
- The conditional block of the first **`if`** statement is followed by the keyword **`else`** and a colon. If the condition in the **`if`** statement is **`True`**, then Python skips the indented block following the **`else`** statement on line 11. If the condition is **`False`**, then Python executes the indented block following the **`else`** statement.
- The **`input`** function appears again in line 13. This time, the string returned by the function is what we want, so no conversion is necessary.
- The notation `!=` in line 14 means "is not equal to." You can also use the keyword **`not`** to negate a Boolean value, so we could have written **`if not`** `(userAgain=='y')`.
- Sometimes a loop should terminate prematurely (before the condition in its **`for`** or **`while`** statement is satisfied). In line 15, the keyword **`break`** instructs Python to exit the indented block of the **`for`** loop and proceed from line 16, regardless of whether the loop is done.
- Line 17 starts a three-way branch whose behavior depends on how the loop terminated. This is accomplished using the **`elif`** statement, short for "else if." There can be as many **`elif`** statements as you like. Python will drop down from the opening **`if`** statement through the **`elif`** statements

[9] See footnote 8.

until it finds one that evaluates to **True** or it reaches an **else** statement. In any circumstance, only one of lines 18, 20, or 22 will be executed.

In short,

> *A branch construction starts with an* **if** *statement. This can be followed by a single block of indented code or multiple blocks separated by* **elif** *or* **else** *statements.*

Include an **else** statement when you want *something* to happen, even when none of the **if** or **elif** statements are satisfied.

The code listed above used the relation operators <, ==, and `!=` to generate Boolean (**True** or **False**) values. Other operators that return Boolean values include >, <=, and >=. You can generate more complex conditionals with the Boolean operators **and**, **or**, and **not**.

3.4.2 Testing equality of floats

Numerical comparisons involving floating point numbers can be dangerous. You are unlikely to ask Python whether two floats representing physical quantities are equal via `a==b`. (Physical effects such as thermal motion, quantum fluctuation, and chaos ensure that this is never the case.) However, consider the following loop:

```
m, k = 0.0, 0.3
while m != k:
    print(m)
    m = m + 0.1
```

It looks harmless enough, but this loop never terminates! On the third iteration, the computer's internal representation of `m` is almost, but not quite, the same as its representation of `k`.[10] Python continues printing indefinitely because `m` is *always* not equal to `k`.

The cure for this particular loop is to use an inequality: `while m<=k: ...` However, you may run into a related problem if you test for equality, thinking that two variables are integers when, in fact, one or both have been converted to floats.

> *Never compare floats with* == *or* `!=`*. Compare integers, or use inequalities.*

If you need to compare two floats, you should probably compare their difference to within some reasonable tolerance appropriate for the problem. NumPy also provides a convenient function to test for "reasonable agreement" between floats, and even allows you to define what is reasonable: See `help(np.isclose)`.

T2 **On truth** Python defines **True** rather broadly. Any nonzero numerical value evaluates to **True** upon conversion to type `bool`, as does any nonempty list, string, tuple, or array. The following expressions evaluate to **False**: **False**, **None**, `[]`, `()`, `{}`, `0`, `0.0`, `0j`, `""`, `''`. Almost everything else evaluates to **True**.

---

[10] T2 The binary representation of 1/10 is a repeating decimal: 0.0001100110011.... This cannot be accurately represented with a finite number of bits. Thus, due to roundoff error, `3*0.1` will not evaluate to exactly the same floating point number as `0.3`.

3.5 NESTING

In many cases, we may wish to embed one **for** loop inside another **for** loop, embed a **for** loop within a **while** loop, or create some other combination. For example, when dealing with probability, we sometimes wish to create an array A where the value of A[i,k] depends on the indices i and k. We can do this with a code like the following:

```
# nesting.py
rows = 3
columns = 4
A = np.zeros( (rows, columns) )
for i in range(rows):
    for k in range(columns):
        A[i, k] = i**2 + k**3
```

Notice that

- Before the loops, the code creates an array with **np.zeros**. NumPy does not build arrays on the fly. It sets aside a block in the computer's memory to store the array, so it needs to know how big that block should be.[11] Thus, we create an array, and then fill it up with the values we want.
- Python's **range** function is used to iterate over values of i and k. It is similar to **np.arange**, but, instead of an array, it creates an object that generates values as they are needed.[12]

Use **np.arange** *when you need an array. Use* **range** *in* **for** *loops.*

- Line 7 sits inside *two* **for** loops, so it is executed $3 \times 4 = 12$ times. One loop inside another is called a **nested loop**. Note that **if** statements and **while** loops can also be nested.
- Line 7 must be executed inside the inner loop, so it is indented twice.
- Descriptive variable names and Python's forced use of indentation make comments nearly superfluous in this code fragment. A reader might only wonder why you are computing $A_{i+1,k+1} = i^2 + k^3$ and not some other function.

[11] You can later resize the array if you need to, but this usually indicates poor planning. If you need a data structure with a variable size, you might use a Python list instead.

[12] T2 The **range** function in Python 3 is equivalent to the xrange function in Python 2. In Python 2, **range** creates an actual list of integers. It will work in **for** loops, but xrange is often preferable, especially for large loops. There is no function called xrange in Python 3.

CHAPTER 4

Data In, Results Out

Computers in the future may... perhaps only weigh 1.5 tons.
— *Popular Mechanics* magazine, 1949

Most data sets are too big to enter by hand, often because they were generated by automated instruments. You need to know how to bring such a data set into your Python computing session (*import* it). You will also want to save your own work to files (*export* it) so that you do not have to repeat complex calculations. Python offers simple and efficient tools for reading and writing files.

Also, most results are too complex to grasp if presented as tables of numbers. You need to know how to present results in a graphical form that humans (including you) can understand. The PyPlot module provides an extensive collection of resources for visualizing data sets.[1]

In this chapter, you will learn to

- Load data from a file;
- Save data to a file; and
- Create plots from a data set.

4.1 IMPORTING DATA

Much scientific work involves collections of experimental data, or *data sets*. In order to crunch a data set, Python must first import it. Many data sets are available in plain text files, including

- *Comma-separated value* (`.csv`) files: Each line of the file represents a row of an array, with entries in that row separated by commas.
- *Tab-separated value* (`.tsv`) files: Each line represents a row of an array, with entries in that row separated by tabs, spaces, or some combination of the two. Spacing does not have to be consistent. Python treats any amount of whitespace between entries as a single separator.

The file extensions `.txt` and `.dat` are also used for data files with comma or whitespace separation. Python has built-in support for reading and writing such files, and NumPy includes additional tools for loading and saving array data to files.

The SciPy library provides a module called **`scipy.io`** that will allow you to read and write data in a variety of formats, including MATLAB files, IDL files, and `.wav` sound files. See `docs.scipy.org`.

---

[1] [T2] All of the data processing described in this chapter—and much more—can be accomplished with the `pandas` module. Though it falls outside the scope of this tutorial, `pandas` provides an extensive suite of tools for working with large data sets. It combines the functionality of NumPy, PyPlot, spreadsheets, and databases into a single interface. The `pandas` library is part of the Anaconda distribution. See `pandas.pydata.org`.

4.1.1 Obtaining data

Before you can import a data set, you must obtain the file of interest and place it in a location where Python can find it. It is good practice to keep all of the files associated with a project—data, scripts, supporting information, and the reports you produce—in a single folder (also called a **directory**).

You can tell Spyder to use a particular folder by setting the **global working directory** in the Preferences menu.[2] Select `Global Working Directory` from the choices on the left. In the box labeled `Startup` select "`the following directory:`". Then choose a folder[3] and click APPLY then OK. Close Spyder and relaunch. Now each time you launch Spyder, it will start in this folder. This is also where Spyder will attempt to load and save all data files by default. The IPython command `%run myfile.py` will also look in this folder.

It is possible to switch to a different working directory during a session in Spyder. We recommend keeping things simple by working in a single folder.

Once you have set up your working directory, you can place data files into it. To download the data sets described in this tutorial, go to

`press.princeton.edu/titles/11349.html`

Here, you will find a link to a zipped file that contains a collection of data sets.[4] Follow the instructions to save the zipped file to your computer. It will probably be saved into your `Downloads` folder unless you specify a different destination. Double-click on the file to unzip it. This will create a new folder called `PMLSdata` in the current folder. You will find all of the data sets described in this tutorial, plus several more. Move `PMLSdata` to a permanent location. If you place `PMLSdata` in your working directory, you will be able to access it easily from within Spyder.

You do not have to limit yourself to the data sets described here. Someday, you'll use data from another source. You may have an instrument in the lab that creates it, or you may get it from a public repository or a coworker. Scientific publications that contain quantitative data in the form of graphs are another source. A graph is a visual representation of a set of numbers, and can be converted back to numbers with a special purpose application. Applications that convert graphs to numbers include

- Engauge Digitizer (`markummitchell.github.io/engauge-digitizer/`),
- Plot Digitizer (`plotdigitizer.sourceforge.net`),
- GraphClick (`www.arizona-software.ch/graphclick/download.html`),
- DataThief (`www.softpedia.com/get/Science-CAD/DataThief.shtml`), and
- GetData (`www.getdata-graph-digitizer.com`).

4.1.2 Bringing data into Python

After you have obtained the data sets, find the entry called `01HIVseries`. This folder contains a file called `HIVseries.csv`. There is also an information file called `README.txt` that describes the data set. Copy these two files into your working directory.

Arrays

You are now ready to launch Python and load the data set. First, inspect the data using the Editor

---

[2] One way to access the Preferences is by clicking the wrench icon.

[3] We recommend a generic name, like `scratch` or `curr`. After you are done with a project, you can archive the files to a different folder, and start the next one in the original scratch folder.

[4] Resources mentioned here are also maintained at a mirror site: `physicalmodelingwithpython.blogspot.com`.

by typing `%edit HIVseries.csv`. (You can also open files in the Editor by using the `File>Open` menu option or by clicking in the Editor window and using `<Cmd-O>`, but you may need to tell Spyder to look for "all files" instead of just Python files.) You should see 16 lines of data, each containing two numbers separated by a comma. You can open the information file `README.txt` to find out what the numbers mean.

To load the data into a NumPy array, you can use the **`np.loadtxt`** command. This command attempts to read a text file and transform the data into an array. By default, NumPy assumes that data entries are separated by spaces and/or tabs. To load a `.csv` file, you must instruct **`np.loadtxt`** to use a comma as the delimiter. A look at the help file (try typing **`np.loadtxt?`**) shows that you can do this by including a *keyword argument* called `delimiter`. Keywords are optional arguments you can pass to a function to modify its behavior (see Section 1.4.4).[5]

To load the data, use the command

```
data_set = np.loadtxt("HIVseries.csv", delimiter=',')
```

Notice how this command is used:

1. The function returns an array. We assign the variable name `data_set` to this array.
2. We must give **`np.loadtxt`** the file name as a string.
3. We must specify the delimiter as a string, in this case `','`.

Python will look for the file in the current working directory. In Spyder, you can see the working directory in a text box in the upper-right corner of the screen. You can use the IPython magic command `%pwd` to print the name of the directory to the screen. If the file fails to load, either it is located in the wrong folder or Spyder is looking in the wrong folder. If Spyder is no longer in its default working directory, you may wish to restart Spyder and try again. If the file is in the wrong location, move it to the working directory and try to load it again. An alternative to moving the file is to specify its complete **path** to the file. For example, on a Mac where the `PMLSdata` folder is still in the `Downloads` folder, you could write[6]

```
data_file = "/Users/username/Downloads/PMLSdata/01HIVseries/HIVseries.csv"
data_set = np.loadtxt(data_file, delimiter=',')
```

Path names are formatted differently in Windows:

```
data_file = r"c:\windows\Downloads\PMLSdata\01HIVseries\HIVseries.csv"
```

The modifier `r` before the string tells Python that this is a *raw string*. In raw strings, backslashes are interpreted literally, not as an escape character.

If you are working with multiple files outside the current working directory, specifying the complete path can lead to a lot of typing. Also, you might move the data to a different folder at some point in the future. For these reasons, it is often convenient to "define once, reuse often" in scripts. For example, with the following two commands near the top of the script, you can easily load multiple data files from the same folder:

```
home_dir = "/Users/username/"
data_dir = home_dir + "Downloads/PMLSdata/01HIVseries/"
data_set = np.loadtxt(data_dir + "HIVseries.csv", delimiter=',')
```

---

[5] T2 By default, **`np.loadtxt`** will ignore any line in the file that starts with #. You can adjust which header lines are discarded by changing the default values of the keyword arguments `comments` and `skiprows`.

[6] Substitute your own username in the place indicated. If you do not know a file's full path, find the file in Finder or File Explorer and drag it into the IPython console. You can then copy and paste the file's path into your code.

Later, if you work on a different computer (where `home_dir` will change) or move the data on your own computer (where `data_dir` will change), you will only have to modify one or two lines to get your code running again.

Once you have successfully imported the data, look at the array. It has two columns. You can see its size and its entries in the Variable Explorer, or by using the IPython Console. There is no description of what the numbers mean. For this, you need to consult `README.txt`.

T2 Other kinds of text

The data file you need to load may not be in any conveniently delimited format. One way to handle such files is to process the file line by line. The following will generate the same array as **`np.loadtxt`**, but it can be adapted to more exotic formats.

```
# import_text.py
my_file = open("HIVseries.csv")
temp_data = []
for line in my_file:
    print(line)
    x, y = line.split(',')
    temp_data += [ (float(x), float(y)) ]
my_file.close()
data_set = np.array(temp_data)
```

The second line creates an object that can read the data file.[7] The third line creates an empty list to store the data. (We use a list instead of an array because we do not know how many lines the file contains.) The **`for`** loop uses a special Python construction to process the file one line at a time. `line` is a string that contains the current line of the file. It is updated during each iteration of the loop. Inside the loop, Python displays the line being processed, and then uses the string method `line.split(',')` to break the line into numbers using the comma as a delimiter. Next, it converts the text into numbers and stores the data from each line as a new ordered pair at the end of the list of data points. The loop will terminate once it reaches the end of the file. After the loop exits, we close the file and convert the list of data points to a NumPy array.

The block of code within the **`for`** loop that processes the data file can be modified to extract data from other types of text files.

T2 Direct import from the web

Instead of saving a data set to your computer and then reading that file, you can combine these steps and read a file directly from the web. Try the following:

```
import urllib
web_file = urllib.request.urlopen( "http://www.physics.upenn.edu/biophys/" + \
                                   "PMLS/Datasets/01HIVseries/HIVseries.csv" )
data_set = np.loadtxt(web_file, delimiter=',')
```

After the call to **`urlopen`**, Python can read the data in `web_file` by using **`np.loadtxt`** (or any other method for processing text files), as if the data set were located in a local file.[8]

---

[7] There is no `file` object type in Python 3, but `open('temp.txt')` returns an object similar to a Python 2 `file` object.

[8] T2 In Python 2, the relevant command is **`urllib.urlopen`**.

4.2 EXPORTING DATA

Saving your work is always important when working on a computer. It is even more important when you are working in the IPython console.

Data on the screen is fleeting. Data saved in files is permanent.

If you have been working for hours and finally finish crunching numbers and making plots, great! But the moment you quit Spyder, all that work is gone. Only data and code that you have saved to files will outlast your session.

4.2.1 Scripts

Scripts are a good way to save your work. (See Section 3.3.) If you keep adding lines to a script as you work through a problem, then you will have a record of your work *and* a program that will reconstruct the state of the system right up to the last line of the script. It can build and fill arrays, load data, carry out complicated analyses, create plots, and much more. (If you have been working at the IPython command line, copying and pasting commands from Spyder's "History log" can help turn your interactive session into a reusable script.)

Each time you run a script, Spyder first saves it. Unless you have modified the global working directory in the Preferences, however, the default folder is probably not what you want. You can use `File>Save As...` to specify the destination. Be sure you know where your script is saved. (See Section 4.1.1 and Appendix A.)

You can also store all or part of your script in other files. Because scripts are plain text files, you can easily copy and paste code into a word processor, a homework assignment, your personal coding log, or an e-mail.

4.2.2 Data files

In contrast to scripts, Python does *not* automatically save its state or the values of any variables created during a session. Hence, the importance of writing scripts: They can reconstruct the state from scratch. However, with large data sets and complex analyses, rerunning the script could take a long time. Once you have generated the data you need, you can save it to a file and simply load it the next time you want to use it.

Data to be read by Python

NumPy provides three convenient functions for saving data stored in arrays to a file:

- **`np.save`** – save a single array as a NumPy archive file (not human readable), with extension `.npy`
- **`np.savez`** – save multiple arrays as a single NumPy archive file (not human readable), with extension `.npz`
- **`np.savetxt`** – save a single array as a text file (human readable), with any extension

For example, the following code will store two arrays in five different files:

```
# save_load.py
x = np.linspace(0, 1, 1001)
y = 3*np.sin(x)**3 - np.sin(x)

np.save('x_values', x)
np.save('y_values', y)
np.savetxt('x_values.dat', x)
np.savetxt('y_values.dat', y)
np.savez('xy_values', x_vals=x, y_vals=y)
```

Explore the resulting files, which have been created in the current working directory. Those created by **np.save** and **np.savez** have the extensions `.npy` and `.npz`, respectively, whereas those created by **np.savetxt** have the `.dat` extension we provided. (They contain tab-separated values. To generate comma-separated values, you must use the *delimiter* keyword again.)

You can view any of these files by opening them in a text editor. Do not save any changes to a `.npy` or `.npz` file when you close it—that will corrupt the data. If you need to show your results to someone else, or use them outside of Python, use **np.savetxt**. As you will see, `.npy` and `.npz` files are not easy to read.

To recover the saved data at any point—in the current session or in the future—use **np.load** or **np.loadtxt**, depending on the file type:

```
x2 = np.load('x_values.npy')
y2 = np.loadtxt('y_values.dat')
w = np.load('xy_values.npz')
```

The variables `x2` and `y2` now refer to arrays containing the same data as `x` and `y`. In the third example, `w` is not an array at all; however, our data is stored inside it.[9] If you type `w.keys()`, Python will return a list with two elements: the two strings `'x_vals'` and `'y_vals'`. These are the names we provided as keyword arguments in the strange-looking call to **np.savez** above. We can access our data from `w` by using the keys. Executing the following lines demonstrates that the data have been faithfully saved and loaded:

```
x2 == x
y2 == y
w['x_vals'] == x
w['y_vals'] == y
```

From these examples, you can see that **np.savez** offers a convenient way to save several arrays in a single object. *Be sure you provide descriptive names for your arrays as keyword arguments.* If you do not, NumPy will use the unenlightening names "`arr_0, arr_1, ...`"

Data to be read by humans

Not all the data you generate will be in arrays. Even with array data, sometimes you will need to write information to a text file, and you may want more control over its format than **np.savetxt** offers. (Perhaps the file will be read by a colleague, or by an application other than Python.)

[9] [T2] It is a special object whose contents can be accessed like those of a Python **dictionary**. This tutorial won't discuss dictionaries, but they are a useful data type. See, for example, Guttag, 2016 or Libeskind-Hadas & Bush, 2014.

Recall Section 3.3.4, where snowball impact times were simply displayed in the console. When the session ends, this information will be lost unless you write it to a file. For small amounts of information, you can simply copy from the IPython console (or History log) and paste into a text editor or word processor.

For large data sets, you will want to write data to a file *without* displaying it on the screen. Python's built-in function **open** will allow you to write directly to a file. To illustrate the method, let's display the first ten powers of 2 and 3 *and* store these data for later use. The following script accomplishes both tasks:

```
# print_write.py
my_file = open('power.txt', 'w')
print( " N \t\t2**N\t\t3**N" )                  # Print labels for columns.
print( "---\t\t----\t\t----" )                  # Print separator.
my_file.write( " N \t\t2**N\t\t3**N\n" ) # Write labels to file.
my_file.write( "---\t\t----\t\t----\n" ) # Write separator to file.
#%% Loop over integers from 0 to 10 and print/write results.
for N in range(11):
    print( "{:d}\t\t{:d}\t\t{:d}".format(N, pow(2,N), pow(3,N)) )
    my_file.write( "{:d}\t\t{:d}\t\t{:d}\n".format(N, pow(2,N), pow(3,N)) )
my_file.close()
```

Compare the output displayed to the screen with the content of `power.txt`.

Writing text to a file is similar to printing to the display.

The differences are that

- You must open a file before writing to it: `my_file=open('file.txt','w')`. (The `'w'` option means this file will be opened for writing. This erases any existing file of the same name *without* any warning or request for confirmation.)
- You must explicitly tell Python where you want a line to start and end. Wherever you have typed `'\n'`, Python will insert a new line. (Similarly, wherever you have typed `'\t'`, Python will insert a tab.)
- When you are done writing, you should close a file with the command `my_file.close()`.

Python is a powerful tool for reading, writing, and manipulating files, but we will not explore these capabilities any further. You can already accomplish quite a lot using the basic commands above in combination with the string formatting methods of Section 2.3.

What should you name your file? Python will accept any valid string you type, but your operating system may not be as flexible. It is good practice to restrict filenames to letters, numbers, underscore (_), hyphen (-), and period.

4.3 VISUALIZING DATA

With data in hand, we are ready for graphics. Python has no built-in graphing functions. Almost all of the graphing tools we discuss come from the PyPlot module, which is part of the larger Matplotlib module. To gain access to these functions, you must import the PyPlot module:

```
import matplotlib.pyplot as plt
```

4.3.1 The plot command and its relatives

PyPlot will make an ordinary, two-dimensional graph for you if you supply it with a set of (x, y) pairs. It's up to you to space those points appropriately. Try

```
# simple_plot.py
import numpy as np, matplotlib.pyplot as plt
num_points = 5
x_min, x_max = 0, 4
x_values = np.linspace(x_min, x_max, num_points)
y_values = x_values**2
plt.plot(x_values, y_values)
# plt.show()
```

The last line may be necessary in your Python environment. If no figure window opens, try uncommenting the last line or typing **plt.show**() at the IPython command prompt.

The figure window may appear automatically, but it can still be hard to find. Try hiding your other applications and moving the Spyder window around. Your plot may be behind it. Changing to a different graphics *back end* may help.[10] (See Section A.2.2, page 117.)

The code given above is a basic plotting script. Notice what happens when you execute it:

- A new window appears. (See Section A.2 if it does not.) This is now the *current figure*. Python will name it "Figure 1" if no other figure windows are open.
- Inside the current figure, PyPlot has drawn and labeled some axes. These are now the *current axes*.
- PyPlot has automatically drawn a region of the xy plane that contains all the points you supplied.
- The **plot** function takes the items in its first argument one at a time, combines them with the corresponding items in the second argument to make (x, y) pairs, and joins those points by a solid blue line. This graph is a bit jagged, but you can improve that by specifying a larger value for num_points. Try it.
- With some back ends, the figure window remains "live" until you close it. That is, you can continue to modify the figure by issuing additional commands at the IPython prompt. Other back ends may force you to close a figure before you can type any more commands.[11] See Section A.2.2 for instructions on how to change the back end.
- Changing the data in y_values will not automatically update your plot. You must issue another **plot** command to see such changes.
- You can close a figure window manually in the usual way (clicking on a red dot or "X" in the corner of the figure window), or from the IPython command prompt with the command **plt.close**(). You can close all open figure windows with the command **plt.close**('all').

The two arrays that you supply to **plt.plot** must be of exactly the same length. That may sound obvious, but it's surprising how often you can get just one extra element in one of them. If you're not sure exactly how many elements you'll get from a=**np.arange**(0,26,0.13), just take a few seconds to type **len**(a) at the IPython console prompt and find out. The function **len** is Python's built-in function for counting the number of elements in a data structure. You can also use **np.shape**(a), to

[10] In the blog that accompanies this book, we describe a way to raise a figure window to the foreground using the Qt back end. (See press.princeton.edu/titles/11349.html).

[11] Calling the function **plt.ion**() from the IPython command line will turn interactive plotting on, if possible.

ask an array about its shape, or request the attribute `a.shape`.

Get in the habit of checking the lengths of your arrays.

Python's **assert** command provides a useful test (Section 3.3.3, page 38). For the example above, you could insert a single statement before the plot command:

```
assert len(x_values) == len(y_values), \
    "Length-mismatch: {:d} versus {:d}".format(len(x_values), len(y_values))
plt.plot(x_values, y_values)
```

PyPlot will issue its own error message if you try to plot lists of unequal lengths, but an **assert** statement provides you with more information about the problem. To see the difference, rewrite the script to change the length of `x_values` after defining `y_values` and run the script with and without the **assert** statement.

You can avoid many of these length mismatch errors by adhering to the principle of "Define once, and reuse often." For example, if the length of every array is determined by the single parameter `num_points`, then they will all have the same length.

Plotting options

Every visual aspect of a graph can be changed from its default. You can specify these when you create a plot; most can also be changed later. Unless you only want to look at a rough plot of some data once and then discard the graph, you should write a script to generate your plot. Otherwise, you will find yourself retyping a lot of commands just to change a label or the color of a single curve in the plot. Here are some of the more common adjustments you can make:

- You can change the color of a curve when you first plot it by adding an optional argument: `plt.plot(x_values,y_values,'r')` produces a solid red line. Other line styles include
  - `'b'` (blue), `'k'` (black), `'g'` (green), and so on.
  - `':'` (join data points with a dotted line), `'--'` (dashed line), `'-'` (solid line), and so on. Remember that red, blue, and other colors look similar on a grayscale printout. Dotted and dashed lines are much easier to distinguish.
  - `'.'` (draw a small dot at each data point), `'o'` (open circle), and so on.

  To some extent, options can be combined: For example, `plt.plot(x_values,y_values,'r--o')` plots a red dashed line with red circles at each point. Python's default is to draw a solid line. If you specify a point style, but no line style, no line will be drawn.
- You can change Python's choice of plot region even *after* creating the plot. Use the command `plt.xlim(1,6)` to display the region $1 \leq x \leq 6$, using the default for y. Similarly, `plt.ylim` adjusts the vertical region. Both functions will also accept keyword arguments,[12] or a list, tuple, or array containing the lowest and highest values to display on the axis: `plt.xlim(xmax=6, xmin=1)`, `plt.xlim([1,6])`, `plt.xlim((1,6))`.
- The command `plt.axis('tight')` effectively issues **xlim** and **ylim** commands to make the axes just fit the range of the data, without extra space around it. (Note the spelling: `plt.axis`, with an 'i', modifies the current axes; `plt.axes`, with an 'e', creates new axes.)
- PyPlot determines the height and width of the figure and then scales the x and y axes by different amounts to make everything fit into the figure. You may not like the distortion of your figure that

[12] Keyword arguments will be discussed in Section 6.1.3.

arises from scaling x and y independently. The command `plt.axis('equal')` forces each axis to use the same scaling; that is, equal coordinate distances give equal distances in the plot. Without this option, a circle may come out looking like an ellipse. The command `plt.axis('square')` does the same thing, but moreover adjusts the coordinate limits so that the graph is actually square in shape.

The documentation on `plt.plot` is quite helpful. Type `help(plt.plot)` or `plt.plot?` at the command prompt to read it. It describes the available marker and line styles, describes other adjustable parameters, and provides several examples.

Some of the commands just mentioned illustrate common themes: Plot attributes can be set by passing strings as arguments to `plt.plot`. Some must appear in a specific position in the list of arguments, like `'r'`. Alternatively, any attribute can be modified by pairing a keyword with the desired value, as in `linestyle='r--'` or its abbreviated form `ls='r--'`.

If you don't want to join your plotted points with a line, there is a separate function called `plt.scatter` that can be called instead of using the options listed above. It gives you more control over the size, style, and color of individual symbols than `plt.plot`.

Log axes

If you'd prefer a logarithmic vertical axis, use `plt.semilogy` in place of `plt.plot`. You can also call `plt.semilogy()` after creating a plot. For a logarithmic horizontal axis, use `plt.semilogx`; for a log-log plot, use `plt.loglog`. In the two latter cases, you may want to evaluate your function at a set of values that are uniformly spaced on a logarithmic scale. See `help(np.logspace)` for a function that creates such arrays.

4.3.2 Manipulate and embellish

There is no limit to the time you can spend improving your graphs. Here we introduce a few more useful adjustments. You can locate many others by using your favorite search engine. (Try searching the web for "`python graph triangle symbol`".) Just remember: When you close a figure, your graph is gone. However, if you write a script to construct the figure and add all of your embellishments, you can easily reconstruct it later.

The commands below may look strange at first, but you will get used to them with practice, especially if you record the ones you use in your coding log (see Section 1.2.4).

You see a plot as an image on your screen, but to Python, a plot is an *object* with many associated *attributes*. The object has *methods* for modifying those attributes. To adjust the properties of an existing figure, first assign a variable to the plot object:

```
# graph_modifications.py
ax = plt.gca()
```

In this command, `plt.gca()` (get current axes) returns an object that controls many of the attributes of the plot in the current figure. It is called an `Axes` object, and it includes the axes you see, tick marks, labels, plot data like curves and symbols, and much more. An `Axes` object is fairly complex, with more than 450 attributes and methods! However, you can accomplish a lot by using only a few of these. The commands fall into two general categories: *getting* data from the object and *setting* values of attributes that you want to adjust. These operations have fairly descriptive names, like `ax.get_xticks()` and `ax.set_xlabel()`. Many of these operations can also be accomplished with functions from the PyPlot module itself.

Here are a few useful operations:[13]

Title: You can add a title with `ax.set_title("My first plot")`. If you don't like the default font, you can change it by using keyword arguments:

```
ax.set_title("My first plot", size=24, weight='bold')
```

Alternatively, you may use

```
plt.title("My first plot", size=24, weight='bold')
```

Axis labels: You can—and should—label the axes with

```
ax.set_xlabel("speed")
ax.set_ylabel("kinetic energy")
```

or

```
plt.xlabel("speed")
plt.ylabel("kinetic energy")
```

Even better, you can—and should—include units: `ax.set_xlabel("speed [um/s]")`. Again, "`um`" is an informal abbreviation for micrometer. But it's almost as easy to write `$\\mu$m` instead, which will produce the standard abbreviation μm. (See Section 2.3.) Even if you are reporting a dimensionless quantity, like concentration relative to its value at time zero, help your readers by stating "`c/c(0) [unitless]`" or "`concentration [a.u.]`."

Tick labels: You can change the font and size of the numbers labeling the tick marks along each axis by first creating the plot, then giving the commands

```
ax.set_xticklabels(ax.get_xticks(), family='monospace', fontsize=10)
ax.set_yticklabels(ax.get_yticks(), family='monospace', fontsize=10)
```

Line style: You can change the attributes of the plotted lines after the initial plot command has been executed. First, we need to gain access to the objects Python uses to draw the lines, then set their properties using PyPlot's **`plt.setp`** command:

```
lines = ax.get_lines()      # Lines is a list of line objects.
# Make the first line thick, dashed, and red.
plt.setp(lines[0], linestyle='--', linewidth=3, color='r')
```

Legend: You can add a **legend**—a descriptive label for each line. You can do this in three ways: Give each line a label when it is created by using the *`label`* keyword, use the **`set_label`** method of a line object, or use the **`legend`** method of an `Axes` object.

```
# Use "label" keyword to set labels when plotting.
plt.plot(x_values, y_values, label="Population 1")
plt.plot(x_values, x_values**3, label="Population 2")
plt.legend()                        # Display legend in plot.
```

[13] The following commands were tested on the `Qt5Agg` back end. They may not work correctly with all graphics back ends.

```
# Use line objects to set labels after plotting.
ax = plt.gca()
lines = ax.get_lines()
lines[0].set_label("Infected Population")
lines[1].set_label("Cured Population")
ax.legend()                               # Display legend in plot.
```

```
# Use Axes object to set labels after plotting.
ax.legend( ("Healthy", "Recovered") )
```

No legend will appear in your plot until you call `ax.legend()` or `plt.legend()`.

These examples show that you can add a lot of useful information to a plot and make it visually appealing. You can also see that this involves a lot of typing. Once you have your plot looking the way you want—or once you get a single aspect of it looking the way you want—copy the commands you used into a script. At the end of the process, you will have a file that can generate the same wonderful graphics at any time on any computer that runs Python.

Your Turn 4A

Start with the code at the beginning of Section 4.3.1 and write a script to produce a figure with a smooth, thick red line. Add labels for the axes, a title, and a legend. Make the text big enough to read easily.

Replacing curves

Sometimes you do not need a new figure—you may simply want to replot a modified version of your data. You can remove all data and formatting from the current figure by calling `plt.cla()`, PyPlot's clear axes command, and then calling `plt.plot` again. However, this removes all of your formatting.

To replace a curve without clearing all of your formatting, you can update the data of individual line objects within the plot:

```
plt.plot(x, 3*x)                  # Plot two lines.
plt.plot(x, x**3)
plt.xlabel('Time [s]')            # Label axes.
plt.ylabel('Position [cm]')
ax = plt.gca()                    # Assign a name to the current Axes object.
lines = ax.get_lines()            # Assign a name to its list of Line objects.
lines[1].set_data(x, x**2)        # Replace plot data of second line.
```

The shape of one curve will change, but all other properties of the plot will be preserved.

4.3.3 T2 More about figures and their axes

PyPlot uses a hierarchy of objects to create a plot. The most important of these for our purposes are the `Figure` and `Axes` objects.

Calling `plt.figure()` creates and returns a `Figure` object. A common side effect of the function is to open a figure window on your screen. If you have several figure windows open, the active window—or the most recently active one—is the "current" figure and will be returned by `plt.gcf()` when no argument is given.

A `Figure` object is not capable of plotting a function; it simply contains and manages all of the elements of a plot. To actually make a graph, we need the second type of object: an `Axes` object.

An `Axes` object is always associated with a `Figure` object. We usually create one by calling PyPlot's convenient **`plt.plot`** function. This function adds an `Axes` object to the current figure, uses it to graph the data we provide, and returns a list of the line objects that were added to the plot. The `Axes` object possesses all of the attributes and methods needed to draw a graph. Many of these methods create and manage yet more objects. You can access these subsidiary objects—things such as lines (associated with a `Line2D` object), axis labels (associated with Python strings), tick marks (associated with a NumPy `ndarray`), and so on—through the parent `Axes` object. Many of *these* manage their own subset of objects necessary to perform their function. You can assign a variable to any of these objects and control its behavior through the variable, as demonstrated in the `ax.get_lines()` example above. This is often more convenient than accessing a property through the parent `Axes` object.

A full discussion of this topic lies well beyond the scope of this tutorial. Fortunately, the PyPlot module provides convenient tools for managing this hierarchy of objects without requiring much understanding of the details. Commands like **`plt.plot`**, **`plt.xlim`**, and **`plt.legend`** are convenience functions that use the methods of `Figure` and `Axes` objects to carry out common operations in making plots.

You can do a lot with PyPlot; however, to get the most control over your plots, you may need to delve into the details of Matplotlib's `Figure` and `Axes` objects. Exploring the objects returned by **`plt.gcf()`** and **`plt.gca()`** is a good place to start. The official documentation is at `matplotlib.org`.

4.3.4 T2 Error bars

To make a graph with error bars, use

```
plt.errorbar(x_values, y_values, yerr=y_errors, xerr=x_errors)
```

This function doesn't add error bars to an existing plot; rather, it creates a new plot (or a line within an existing plot) with error bars. Its syntax is similar to **`plt.plot`**, but it accepts additional optional arguments. Point `n` will be augmented by error bars that extend coordinate distance `y_errors[n]` above and below the point, and coordinate distance `x_errors[n]` to the left and right of the point. Both error arguments are optional. If you supply only one, there will be no error bars in the other direction. If you do not supply either, **`plt.errorbar`** behaves like **`plt.plot`**.

4.3.5 3D graphs

Sometimes a graph consists of points or a curve in a three-dimensional space. To make 3D plots, we must import an additional tool from one of the Matplotlib modules:

```
# line3d.py
from mpl_toolkits.mplot3d import Axes3D      # Get 3D plotting tool.
```

We must also interact with PyPlot in a different way. The following script will generate a helix:

```
fig = plt.figure()                      # Create a new figure.
ax = Axes3D(fig)                        # Create 3D plotter attached to figure.
t = np.linspace(0, 5*np.pi, 501)        # Define parameter for parametric plot.
ax.plot(np.cos(t), np.sin(t), t)        # Draw 3D plot.
```

Notice that this time we are using the **plot** method attached to a specific object. We created an **Axes3D** object and can now use its methods to generate three-dimensional plots. The method ax.**plot** accepts three arrays, whose corresponding entries are interpreted as the x, y, and z coordinates of each point to be drawn. The object generated by **Axes3D**(fig) can also create surface plots, three-dimensional contour plots, three-dimensional histograms, and much more.

Your screen is two-dimensional. If you ask for a "3D" plot, what you actually get must be a two-dimensional projection—what a camera would see looking at your plot from some outside **viewpoint**. Switching to a different viewpoint may make it easier to see what's going on. You may be able to click and drag within your plot to change the viewpoint. (This depends on the graphics back end you are using. See Section A.2.) If this is not possible, you can change the viewpoint from the IPython command prompt or in a script with the command ax.**view_init**(*elev*=30,*azim*=30). You specify the viewpoint by giving the elevation and azimuth in degrees—not radians.

4.3.6 Multiple plots

Multiple plots on the same axes

You probably noticed in the previous examples that every new **plt.plot** command adds a new curve to an existing figure, if one is already open.

By default, a new plot is added to the current axes of the active figure.

You can also create several curves on the same axes with a single plot command. Try

```
x = np.linspace(0, 1, 51)
y1 = np.exp(x)
y2 = x**2
plt.plot(x, y1, x, y2)
```

This example shows that you can give **plot** more than one set of (x, y) pairs. Python will choose different colors for each curve, or you can specify them manually:

```
plt.plot(x, y1, 'r', x, y2, 'ko')
```

A third way to draw multiple curves on one set of axes is to give **plt.plot** an ordinary vector of x values, but a two-dimensional array of y values. Each column of y will be used to draw a separate curve using a common set of x values. This approach can be useful when you wish to look at a function for several values of a parameter.

Your Turn 4B

Try the following code:

```
num_curves = 3
x = np.linspace(0, 1, 51)
y = np.zeros( (x.size, num_curves) )
for n in range(num_curves):
    y[:, n] = np.sin((n+1) * x * 2 * np.pi)
plt.plot(x, y)
```

Add a legend explaining which curve is which, then use the methods of Section 4.3.2 to embellish the plot.

Multiple plot windows

If you want two or more separate plots open, use **plt.figure**() to create a new figure and make it active, then use **plt.plot** to generate the second plot. If you don't supply an argument, **plt.figure**() chooses a figure number that is not in use yet. You can also assign a name you choose yourself. For example, **plt.figure**('Joe') will create a figure named "Joe."[14] The next **plt.plot** command will draw its output in the current figure, with no effect on other figures.

If you are generating several plots in a session, you will probably want to create a new figure each time by using **plt.figure**(). However, these plots can use a fair amount of your computing resources, and finding the graph you want among a dozen figure windows is not always simple. So once you are finished with a figure, you can close it. You can use **plt.close**() to close the current figure, or you can close a figure by name: **plt.close**('Joe').

*You can close **all** open figures with* **plt.close**('all').

Include this command near the top of a script to ensure that all figures come from the latest run.

4.3.7 Subplots

You may wish to place multiple plots side by side in a single figure window for comparison. In Python jargon, you'd like multiple *axes* occupying the same *figure.* The function **plt.subplot**(M,N,p) divides the current figure window into a grid of cells with M rows and N columns, then makes cell number p the active one. Here, p must be an integer between 1 and M*N. (This is one case where Python does *not* start counting from 0.) Any graphing command issued now will affect cell number p. Try this example:

```
# subplots.py
from numpy.random import random
t = np.linspace(0, 1, 101)
plt.figure()
plt.subplot(2, 2, 1); plt.hist(random(20))                    # Upper left
plt.subplot(2, 2, 2); plt.plot(t, t**2, t, t**3 - t)          # Upper right
plt.subplot(2, 2, 3); plt.plot(random(20), random(20), 'r*')  # Lower left
plt.subplot(2, 2, 4); plt.plot(t*np.cos(10*t), t*np.sin(10*t)) # Lower right
```

Each **plt.subplot** command selects a subregion of the figure. The subsequent **plt.plot** command sets up axes and draws a plot in the current subplot region. If your subplots do not fit nicely into the figure window or overlap one another, you can resize the figure window or try the command **plt.tight_layout**().

Make sure that all your **plt.subplot** commands for a particular figure use the same values for N and M. If you change either, all existing subplots will be erased. For more freedom in placing subplots and insets, see **help**(**plt.axes**).

4.3.8 Saving figures

All of your figures will disappear when you end your session in Spyder. If you used a script to create and adjust your plots, you can recreate them later, but you may also want to use plots in other

[14] Or make it the current figure if it already exists. Either way, **plt.gca**() returns the current Axes object of the current figure, so you can use **plt.figure**('Joe') to select "Joe," then **plt.gca**() to adjust a plot's attributes, even if there are multiple figures open.

applications or print them out. Saving a plot to a graphics file for such purposes is straightforward. The figure window has a `Save` icon that will open a dialog allowing you to save the current figure in a variety of formats. The default is `.png`, but most Python distributions will also allow you to save the figure in `.pdf`, `.jpg`, `.eps`, or `.tif` format. To get the complete list, consult the figure:

```
fig = plt.gcf()                           # Get current figure object.
fig.canvas.get_supported_filetypes()
```

All you need to do is give the figure a name with the appropriate extension; PyPlot will create a graphics file of the type you requested.

You can also save a figure from the IPython command prompt, or from within a script, by using the command **plt.savefig**. For example, to save the current figure as a `.pdf` file, type

```
plt.savefig("greatest_figure_ever.pdf")
```

The plot will be saved in your current working directory. (You can force Python to save in a different directory by giving the complete path, as discussed in Section 4.1.2, page 49.)

4.3.9 T2 Using figures in other applications

If you save a figure as an `.eps` or `.svg` file, you can open and modify it in a *vector-graphics* application such as Inkscape or Xfig (both freeware), or a commercial alternative.[15] Text such as the title or axis labels may be rendered as "outlines," which can make it difficult to edit in another application. If you encounter this problem, you can instruct Python to save text as "type" in an `.svg` file. Before saving the figure, modify the parameter that controls how fonts are saved in `.svg` figures:

```
import matplotlib
matplotlib.rcParams['svg.fonttype'] = 'none'
```

Then, save your figure as an `.svg` file.

Some applications won't read `.svg` files prepared in this way. An alternative that still gives "type" is to use

```
import matplotlib
matplotlib.rcParams['pdf.fonttype'] = 42
```

Then, save your figure as a `.pdf` file, and edit it in another application. For more information about **matplotlib**.`rcParams`, see "Customizing matplotlib" at `matplotlib.org`.

If you are going to use figures in a presentation, you may instead need to save images in a *raster* format (also called **bitmap**), such as `.gif`, `.png`, `.jpg`, or `.tif`. For publication, however, vector art is generally best.

[15] Inkscape: `www.inkscape.org`; Xfig: `www.xfig.org`.

CHAPTER 5

First Computer Lab

These exercises will use many ideas from the previous chapter. Our goals are to

- Develop basic graphing skills;
- Import a data set; and
- Perform a simple fit of a model to data.

5.1 HIV EXAMPLE

Here we explore a model of the **viral load**—the number of virions in the blood of a patient infected with HIV—after the administration of an antiretroviral drug. One model for the viral load predicts that the concentration $V(t)$ of HIV in the blood at time t after the start of treatment will be

$$V(t) = A\exp(-\alpha t) + B\exp(-\beta t). \tag{5.1}$$

The four parameters A, α, B, and β are constants that control the behavior of the model.[1]

In this lab, you will use what you have learned about Python up to this point to generate plots based on the model, import and plot experimental data, and then fit model parameters to the data.

5.1.1 Explore the model

To get started, launch Spyder, import NumPy and PyPlot, and then create an array by typing

```
time = np.linspace(0, 1, 11)
time
```

at the IPython command prompt. Press `<Return/Enter>` after each line. You should see a list of 11 numbers. Now modify these commands to create an array of 101 numbers ranging from 0 to 10 and assign it to the variable `time`.

Next, you will evaluate a compound expression involving this array by using the solution of the viral load model given in Equation 5.1.

The first step is to give the constants in the mathematical equation names you can type, such as `alpha` and `beta` instead of α and β. It is wise to give longer, more descriptive names even to the variables whose names you can type: for example, `time` for t and `viral_load` for $V(t)$. Now, set $B = 0$, and choose some interesting values for A, α, and β. Then, evaluate $V(t)$ by using the following command:

```
viral_load = A * np.exp(-alpha*time) + B * np.exp(-beta*time)
```

You should now have two arrays of the same length, `time` and `viral_load`, so plot them:

[1] Nelson, 2015, Chapter 1 gives details of the model and its solution.

```
plt.plot(time, viral_load)
```

Create a few more plots using different values of the four model parameters.

5.1.2 Fit experimental data

Now let's have a look at some experimental data.

Follow the instructions of Section 4.1 to obtain the data set `01HIVseries`. Copy the files `README.txt`, `HIVseries.csv`, `HIVseries.npy`, and `HIVseries.npz` into your working directory. The `HIVseries` files contain time series data. Read the file `README.txt` for details.

Import the data set into Python by reading `HIVseries.csv` into an array or by loading either `HIVseries.npy` or `HIVseries.npz`. (Remember: The functions **`np.load`** and **`np.loadtxt`** return the data you requested, but you must assign that data to a variable to access it later.) You are free to give the data whatever name you like, but we will refer to the array as `hiv_data`.

If you use **`np.loadtxt`** to import `HIVseries.csv` or **`np.load`** to import `HIVseries.npy`, the data will be contained in a single array. Find the variable you created in the Variable Explorer. It has two columns of data. The first column is the time in days since administration of a treatment to an HIV-positive patient; the second contains the concentration of virus in that patient's blood in arbitrary units.

If you instead use **`np.load`** to import `HIVseries.npz`, the data will be contained in an object that does not show up in the Variable Explorer. (See Section 4.2.2.) You can nevertheless inspect its contents. Type `hiv_data.keys()` to find the names of the arrays stored within. There are two keys, called `'time_in_days'` and `'viral_load'`. You can inspect the corresponding arrays by asking for them by name. You can also assign the data in these arrays to a variable name of your choosing: for example, `concentration = hiv_data['viral_load']`.

Next, we are going to visualize the data. In order to plot the viral load as a function of time, you will need to separate the data into two arrays to pass to **`plt.plot`**. Do that and plot the data points now. Don't join the points by line segments. Instead, make each point a symbol, for example, a small circle or plus sign. Label the axes of your plot. Give it a descriptive title, too. Such embellishments are discussed in Section 4.3.1.

Assignment:

a. *Superimpose the experimental data points on a plot of the function in Equation 5.1. Adjust the four parameters of the model until you can see both the data and the model in your plot.*

The goal is now to tune the four parameters of Equation 5.1 until the model agrees with the data. It is hard to find the right needle in a four-dimensional haystack! We need a more systematic approach than just guessing. Consider the following: Assuming $\beta > \alpha$, how does the trial solution behave at long times? If the data also behave that way, can we use the long-time behavior to determine two of the four unknown constants, then hold them fixed while adjusting the other two?

Even two constants is a lot to adjust by hand, so let's think some more: How does the initial value $V(0)$ depend on the four constant parameters? Can you choose these constants in a way that always gives the correct long-time behavior *and* initial value?

b. *Carry out this analysis so that you have only* **one** *remaining free parameter, which you can adjust fairly easily. Adjust this parameter until you like what you see.*

c. *The latency period of HIV is about ten years. Based on your results, how does the inverse of the T-cell infection rate, $1/\alpha$, compare to the latency period?*

[*Remark:* You probably know that there are black-box software packages that do such "curve fitting" automatically. In this lab, you should do it manually, just to see how the curves respond to changes in the parameters.]

5.2 BACTERIAL EXAMPLE

Now we turn our attention to genetic switching in bacteria. In 1957, A. Novick and M. Weiner studied the production of a protein called beta galactosidase in *E. coli* bacteria after introducing an *inducer* molecule called TMG.[2]

5.2.1 Explore the model

Here are two families of functions that come up in the analysis of the Novick–Weiner experiment:

$$V(t) = 1 - \mathrm{e}^{-t/\tau} \qquad \text{and} \qquad W(t) = A\left(\mathrm{e}^{-t/\tau} - 1 + \frac{t}{\tau}\right). \tag{5.2}$$

The parameters τ and A are constants.

Assignment:

a. *Choose $A = 1$, $\tau = 1$, and plot $W(t)$ for $0 < t < 2$.*

b. *Make several arrays `W1`, `W2`, `W3`, and so on, using different values of τ and A, and plot them simultaneously on the same graph.*

c. *Change the colors and line styles (solid, dashed, and so on) of the lines.*

d. *Add a legend to help a potential reader sort out the curves. Explore some of the other graph options that are available.*

5.2.2 Fit experimental data

Follow the instructions of Section 4.1 to obtain the data set `15novick`. Copy `g149novickA.csv`, `g149novickA.npy`, `g149novickB.csv`, `g149novickB.npy`, and `README.txt` into your working directory. The files contain time series data from a bacterial population in a culture.

Assignment:

a. *Superimpose the experimental data points on a plot of $V(t)$ with multiple curves (see Equation 5.2), similar to the one you made for $W(t)$ before.*

Plot the experimental data points. Don't join the points by line segments; make each point a symbol such as a small circle or plus sign. Label the axes of your plot. Select some reasonable values for the

[2] See Nelson, 2015, Chapter 10 for details of the experiment and model.

parameter τ in the model, and see if you can get a curve that fits the data well. Label the curves, then add a legend to identify which curve is which.

b. *Now try the same thing using the data in* `g149novickB.csv` or `g149novickB.npy`*. This time throw away all the data with time value greater than ten hours, and attempt to fit the remaining data to the family of functions $W(t)$ in Equation 5.2.*

[*Hint:* At large values of t, both the data and the function $W(t)$ become straight lines. Find the slope and the y intercept of the straight line determined by Equation 5.2 in terms of the two unknown quantities A and τ. Next, estimate the slope and y intercept of the straight line determined by the data. From this, figure out some pretty good guesses for the values of A and τ. Then, tweak the values to get a nicer looking fit.]

[*Remark:* Again, you are asked not to use an automatic curve-fitting system. The methods suggested in the Hint will give you a deeper understanding of the math than just accepting whatever a black box spits out.]

CHAPTER 6

More Python Constructions

At its highest level, numerical analysis is a mixture of science, art, and bar-room brawl.

—T. W. Koerner, *The Pleasures of Counting*

The preceding chapters have developed a basic set of techniques for importing, creating, and modeling data sets and visualizing the results. This chapter introduces additional techniques for exploring mathematical models and their predictions:

- Random numbers and Monte Carlo simulations
- Solutions of nonlinear equations of a single variable
- Solutions of linear systems of equations
- Numerical integration of functions
- Numerical solution of ordinary differential equations

In addition, this chapter introduces several new methods for visualizing data, including histograms, surface plots, contour plots, vector field plots, and streamlines.

We start with a discussion of writing your own functions, an invaluable tool for exploring physical models.

6.1 WRITING YOUR OWN FUNCTIONS

Section 3.3.5 introduced a principle:

Don't duplicate. Define once, and reuse often.

In the context of parameters (fixed quantities), this means you should define a parameter's value just once at the start of your code and refer to it by name throughout the rest of the program. But code itself can contain duplications if we wish to do the same (or nearly the same) task many times. Just as with parameter values, you may later realize that something needs to be changed in your code. Changing every instance of a recurring code fragment can be tedious and prone to error. It is better to define a *function* once, then invoke it whenever needed. You may even use the same fragment of code in more than one of your scripts. If each script imports the same externally defined function, then changes that you make once in the function file will apply to all of your scripts.

Functions in Python can do almost anything. There are entire libraries of functions that can carry out mathematical operations, make plots, read and write files, and much more. Your own functions can do all of these things as well. Functions are ideal for writing code once and reusing it often.

6.1.1 Defining functions in Python

A function can be defined at the command prompt or in a file. The following example is a basic template for creating a function in Python:

```
# excerpt from measurements.py
def taxicab(pointA, pointB):
    """
    Taxicab metric for computing distance between points A and B.
        pointA = (x1, y1)
        pointB = (x2, y2)
    Returns |x2-x1| + |y2-y1|. Distances are measured in city blocks.
    """
    interval = abs(pointB[0] - pointA[0]) + abs(pointB[1] - pointA[1])
    return interval
```

The "taxicab metric" determines the billable distance between points "as the cab drives" rather than "as the crow flies." (That is, we use the total distance driven rather than the shortest straight line distance between the two points.)

The function consists of the following elements:

Declaration: The keyword `def` tells Python you are about to define a function. Function names must adhere to the same rules as variable names. It is wise to give your functions descriptive names. Remember: You will only have to define it once, but you will wish to reuse it often. Now is not the time to save typing by giving it a forgettable name like `f`.

Arguments: The name of the function is followed by the names of all its **arguments**: the data it requires to do its calculation.[1] In this case, `taxicab` requires two arguments. If it is called with only one argument or arguments of the wrong type, Python will raise a `TypeError` exception.

Colon: The colon after the list of arguments begins an indented block of code associated with the function. Notice the similarity with `for` and `while` loops and `if` statements. Everything from the colon to the end of the code block will be executed when the function is called.

Docstrings: The text between the pair of triple-quotes (`""" ... """`) is a docstring. It is a special comment Python will use if someone asks for `help` on your function, and it is the standard place to describe what the function does and what arguments are required. Python will not complain if you do not provide a docstring, but someone who uses your code might.

Body: The body of the function is the code that does something useful with the arguments. This example is a simple function, so its body consists of just two lines. More complicated functions can include loops, `if` statements, and calls to other functions, and may extend for many lines.

Return: In Python, a function always returns something to the program that invokes it. If you do not specify a return value, Python will supply an object called `None`. In this example, our function returns a `float` object.

Now, let's look at how a function call works. Enter the function definition for `taxicab` at the command prompt. (You may omit the docstring, but only this one time!) Then, type

---

[1] T2 Arguments in the definition of a function are sometimes called "formal parameters."

```
fare_rate = 0.40            # Fare rate in dollars per city block
start = (1, 2)
stop = (4, 5)
trip_cost = taxicab(start, stop) * fare_rate
```

When the fourth line is executed, `taxicab` behaves like **`np.sqrt`** and other predefined functions (Section 1.4.2, page 13):

- First, Python assigns the variables in the function's argument list to the objects passed as arguments.[2] It binds `pointA` to the same object as `start`, and `pointB` to the same object as `stop`. (See Appendix E for details.)
- Then, Python transfers control to `taxicab`.
- When `taxicab` is finished, Python substitutes its return value into the assignment statement for `trip_cost`.
- Python finishes evaluating the expression and assigns the answer to `trip_cost`.

Although we defined `start` and `stop` as tuples, we can call our new function with any objects that understand what `thing[0]` and `thing[1]` mean: lists, tuples, or NumPy arrays.

Your Turn 6A

> Define a function to compute the straight-line distance between two points in three-dimensional space. Give it a descriptive name and an informative docstring. See what happens when you call it with the wrong number or type of arguments, and ensure that using **`help`** on your function will enable a user to diagnose and resolve the issue.

In some ways, a function is similar to a script. It is a fragment of code that is executed upon request. Unlike a script, a function is a Python object. It can be called by name (once it is imported), and it can be called by another script or function. A function communicates with the calling program by way of its arguments and its return value. After evaluating the function, Python discards all of its local variables, so remember:

If a function performs a calculation, **`return`** *the result.*

You can define functions at the IPython command prompt. However, if you plan to use a function more than once, you should save it in a file. You can place a single function in its own file such as `taxicab.py` and run this file as a script prior to using the function. (Running the file will define the function as if it had been entered at the command prompt.) You can also define a function within the script that uses it, as long as its definition gets executed before it is first called.

If you will be using the same functions in multiple scripts and interactive sessions, you can create a **module** in which you define one or more functions in a single `.py` file. A module is a script that contains a collection of definitions and assignments. It can be imported into your session from the command prompt, or by placing an **`import`** command within a script. You can also import selectively, as described in Section 1.3.

Place `taxicab` and the function you wrote in Your Turn 6A (called, for example, `crow`) in a single file called `measurements.py` in your working directory. Then, type **`import`** `measurements`. You now

---

[2] T2 Arguments passed to a function are sometimes called "actual parameters."

have access to both functions under the names measurements.taxicab and measurements.crow. Type **help**(measurements) and **help**(measurements.taxicab) to see how Python uses the docstrings you provide. Note that if you give a *module* the same name as a *function* it contains, you still have to provide both the module and function name to Python. For example, if you save the taxicab function in a file called taxicab.py and then **import** taxicab, calling taxicab(A,B) will result in an error. To access the function, you must use taxicab.taxicab(A,B).

Modules that you write and wish to use should be located in the same folder as your main script, which is probably the global working directory you specified during setup (as described in Sections 4.1.1 and A.2).[3]

6.1.2 Updating functions

If you have modified a function in a script or in your own module, you will need to instruct Python to use the newest versions. If the function is defined in a script, you can run the script again by using the RUN▶ button (if the script is open in the Editor) or the **%run** command.

If your function is part of a module that you have imported, you will need to restart the IPython kernel or relaunch Spyder for the changes to take effect. Just typing **import** my_module again will *not* update the module—even after calling **%reset**. Alternatively, you can use a function called **reload** to update any module without restarting. This function is part of the **imp** module.[4] For example, if you modify the taxicab function's code in measurements.py after importing measurements as a module, you need to save the file and then type the following commands:[5]

```
from imp import reload
reload(measurements)
```

Again, calling **import** will only cause Python to load a module *if it has not already been imported.*

> *You must save a module and then use* **reload** *to update functions in a module you have already imported.*

Reloading a module is usually necessary only in debugging. Once your functions are working properly, you need only import them once in a script or interactive session.

6.1.3 Arguments, keywords, and defaults

We have already seen that keyword arguments can modify the behavior of functions like **plt.plot**. You can also use keywords and give arguments default values in functions you write. The following example demonstrates both of these techniques:

```
# measurements.py
def distance(pointA, pointB=(0, 0), metric='taxi'):
    """
    Return distance in city blocks between points A and B.
    If metric is 'taxi' (or omitted), use taxicab metric.        5
    Otherwise, use Euclidean distance.
```

[3] T2 If you know about "paths," then you can create folders that Python can access from any working directory.

[4] In Python 2.7, **reload** is a built-in function and does not need to be imported.

[5] You only need to import the **imp** module once. If you make further changes to measurements.py, just type **reload**(measurements).

```
        pointA = (x1, y1)
        pointB = (x2, y2)
    If pointB is omitted, use the origin.
    """
    if metric == 'taxi':
        interval = abs(pointB[0] - pointA[0]) + abs(pointB[1] - pointA[1])
    else:
        interval = np.sqrt( (pointB[0] - pointA[0])**2 \
                          + (pointB[1] - pointA[1])**2 )
    return interval
```

The `distance` function decides how to calculate the distance between points based on the value of `metric`. Both `pointB` and `metric` have default values. If the values for these arguments are passed to the function, it will use them; if not, it will use the defaults. Try

```
distance( (3, 4) )
distance( (3, 4), (1, 2), 'euclid' )
distance( (3, 4), 'euclid' )                              # This is an error.
distance( pointB=(1, 2), metric='normal', pointA=(3, 4) )
```

The arguments to a function must either be in the correct order according to the function definition, as in the first two lines of this example, or be paired with a keyword, as in the last line. The third line results in an error because Python assigns the string literal `'euclid'` to the variable `pointB`.

6.1.4 Return values

A function can take any type or number of arguments—or none at all. The return value of a function is a single object, but that object may be a number, an array, a string, or a collection of objects in a tuple or list. A function will return an object called **`None`** if no other return value is specified.

Suppose that you wish to write a function to rotate a two-dimensional vector. What should the arguments be? What should the function return?

The arguments should include the vector to be rotated and an angle of rotation. The function should return the rotated vector. Here is a function that will accomplish the task:

```
# rotate.py
def rotate_vector(vector, angle):
    """
    Rotate a two-dimensional vector through given angle.
        vector = (x,y)
        angle  = rotation angle in radians (counterclockwise)
    Returns the image of a vector under rotation as a NumPy array.
    """
    rotation_matrix = np.array([[ np.cos(angle), -np.sin(angle) ],
                                [ np.sin(angle),  np.cos(angle) ]])
    return np.dot(rotation_matrix, vector)
```

This implementation of rotation creates a 2×2 matrix and then multiplies this matrix and the vector supplied as the first argument. The function does not modify the contents of `vector`. Instead, **`np.dot`** creates a new array, which is the returned value of the function.

Python allows you to **unpack** compound return values—to assign the individual elements of an object to separate variables. Multiple assignments on a single line, such as `x,y=(1,2)`, are a simple example of unpacking. Python can unpack any iterable object: a tuple, a list, a string, or an array.

There are several ways to unpack an object. The following calls to `rotate_vector` all work, but the components of the rotated vector are split up in different ways:

```
vec = [1, 1]
theta = np.pi/2
r = rotate_vector(vec, theta)
x, y = rotate_vector(vec, theta)
_, z = rotate_vector(vec, theta)
first, *rest = rotate_vector(vec, theta)
```

After executing these commands, you should find that `r` is a NumPy array with two elements. The other variables, `x`, `y`, and `z`, contain the individual components of the rotated vector. The **underscore**, `_`, is a dummy variable whose value is discarded. (Underscore is a special variable, whose value is the result of the most recent command. We are using it to temporarily store values that we do not need.) In the final line, `first` contains the first element of the rotated vector and `rest` is a list that contains everything else.[6] In this case, `rest` contains only one element, but if the function returned an array with 100 elements, `first` would still contain only the first value and `rest` would be a list (*not* a NumPy array) containing the other 99 elements. You could convert it to an array with `np.array(rest)`.

6.1.5 Functional programming

Python offers programmers great flexibility in writing functions. We discuss some of these nuances in Appendix E. However, we strongly endorse the following guidelines for writing your own functions:

1. Pass data to a function only through its arguments.
2. Do not modify the arguments of a function.
3. Return the result of a calculation with a **`return`** statement.

Python allows you to circumvent all of these conventions, but your code will be easier to write, interpret, and debug if you adhere to them. Your functions will accept input (arguments) and produce output (a return value), without side effects.

A **side effect** is any effect on the computer's state other than returning a value. Avoiding side effects is the cornerstone of *functional programming*. Sometimes side effects are useful, but consider the alternatives when writing your own functions.

Unintended side effects on arrays can be particularly troublesome. Python does *not* automatically create a local copy of an array inside a function. As a result, a function can modify the data in an array passed as an argument using array methods—including assignment of individual elements. (See Section 2.2.6, page 24.) Array methods like `x.sort()` and `x.fill(3)`, as well as in-place arithmetic like `x+=1` and `x*=2`, can also modify array data. Such side effects have the potential to speed up your code and reduce memory usage for operations on large arrays. You can **overwrite** the array and modify its elements without making any copies. However, this increase in performance comes at the cost of difficulty in debugging.

If you really do intend for a function to modify the elements of an array, you can still avoid side effects. Create a local copy of the array within the function, operate on the copy, and return the new

[6] This type of unpacking is not available in Python 2.

array.[7] Alternatively, you can create an empty array within the function, fill it with new values, and return the new array. NumPy adheres to this principle: Functions like **np.cos**(x) and **np.exp**(x) return new arrays, with no effect on x. If you wish, you can replace the original array with the new array returned by the function in the main code, rather than inside the function.

The following function illustrates these principles of functional programming:

```
# average.py
def running_average(x):
    """
    Return cumulative average of an array.
    """
    y = np.zeros(len(x))                    # Empty array to store result
    current_sum = 0.0                       # Running sum of elements of x
    for i in range(len(x)):
        current_sum += x[i]                 # Increment sum.
        y[i] = current_sum / (i + 1.0)      # Update running average.
    return y
```

The array being processed is passed as an argument. This array is not modified; the result of the calculation is returned in a new array.

If you need the extra performance and reduced memory usage that come from overwriting an array, use the information in Appendix E to plan your code carefully and avoid unintended consequences. However, try vectorizing your code using NumPy's efficient array operations before you resort to overwriting. Vectorized code will almost always run faster than loops you write yourself in Python.

6.2 RANDOM NUMBERS AND SIMULATION

There are many interesting problems in which we do not have complete knowledge of a system, but we do know the probabilities of the outcomes of simple events. For example, you know the probability of any given outcome in the roll of a single die is $1/6$, but do you know how likely it is that the sum of a roll of 5 dice is less than 13? Rather than work out the combinatorics, you could roll 5 dice many times and determine the probability empirically.

A random number generator makes it possible for a computer to "roll dice" millions of times per second. Thus, you can use a random number generator to simulate a system described by a stochastic model in which the probability distributions of the parameters are known. You can determine the likely behavior of the system, even if you cannot work out the details analytically. Such calculations are often called "Monte Carlo simulations."

6.2.1 Simulating coin flips

The simplest example of a stochastic system is a coin flip. Suppose that you want to simulate flipping a coin 100 times, record the number of heads or tails, and then repeat the whole series of 100 flips N times. This will generate a set of N numbers, each falling in the range of 0 to 100, inclusive.

How do we get Python to flip the coin for us? First, try typing `1>2` at the IPython console prompt, and then `2>1`. You'll see that Python returns a Boolean value of **True** or **False** when it evaluates

[7] You can copy an array using the array's **copy**() method: `y = x.copy()`. After this statement, x and y will be independent arrays with the same data. Note that slicing does *not* create a copy. See Appendix E.

each of these expressions. You can simulate a coin flip by generating a uniformly distributed random number between 0 and 1, then checking to see if it is less than 0.5. If the comparison returns **`True`**, we record heads; if **`False`**, we record tails. Python can also use such values in numeric calculations: It converts **`True`** to 1 and **`False`** to 0.

The module **`np.random`** contains random number generators for several different probability distributions. The function **`np.random.random`** generates numbers from the continuous uniform distribution over the interval $[0, 1)$, and will serve our purposes for this chapter. You should explore the other functions in the **`np.random`** module as well. To save typing, we will import **`np.random.random`** under the nickname `rng`:

```
from numpy.random import random as rng
```

In the future, if you wish to switch to a different random number generator, this is the only line of code you will need to change: Just import a different function under the same nickname `rng`.

To make a series of independent flips, first create an array called `samples` that contains 100 random numbers generated by `rng`. (Consult `help(rng)` for a clue about how to do this easily.) You can convert the random samples to a simulation of coin flips with `flips=(samples<0.5)`. Python applies the comparison item by item and stores the result in `flips`. You can then count the number of heads by using `np.sum(flips)` or the array method `flips.sum()`. Repeat this several times to get a feel for how likely it is that exactly 50 heads occur in 100 flips of a fair coin.

6.2.2 Generating trajectories

We can adapt coin flipping to study random walks, Brownian motion, and a host of other interesting physical and biological systems. Let's create a random walk of 500 steps. Our trajectory will consist of 500 x values and 500 y values. Following good practice (Section 3.3.5), begin with

```
num_steps = 500
```

The idea behind a random walk is that every step is a statistically independent, random event. You know from the preceding section how to get an array containing 500 random binary digits. Make two such arrays called `x_step` and `y_step`. For a coin flip, **`True`** and **`False`** or 1 and 0 were sufficient. For our random walk, however, we need a random string of $+1$ and -1 values.

Your Turn 6B

a. Do a simple algebraic operation on `x_step` that maps each 1 to $+1$ and each 0 to -1. Do the same for `y_step`.
b. Next, convert these arrays into successive positions of the random walker. Consult `help(np.cumsum)` to see why that function is useful.
c. Write a script to make a picture of your random walk, and run it several times.

Try writing a function called `get_trajectory` that will take the single argument `num_steps` and return the two arrays that contain the coordinates of a random walker.

6.3 HISTOGRAMS AND BAR GRAPHS

6.3.1 Creating histograms

A **histogram** is a graphical representation of a discrete probability distribution. Suppose that you wish to check that the random number generator `rng` really gives a uniform distribution. To make a simple histogram, try

```
# histogram.py
from numpy.random import random as rng
data = rng(100)
plt.hist(data)
```

A histogram appears, but a number of things have happened without your supervision. The function **`plt.hist`** has

- Inspected the list of data and determined its range;
- Divided that range into a number of equally spaced **bins**;
- Counted how many elements of `data` fall into each bin;
- Graphed the result as a set of bars; and
- Returned a tuple containing two arrays and a strange-looking item called `<a list of 10 Patch objects>`.

Consulting **`help(plt.hist)`** reveals the content of the return values as well as the keywords and defaults that you can use to control the output. For instance, `bins=10` means that **`plt.hist`** will generate 10 equally spaced bars in the histogram unless you specify another value, and `align='mid'` specifies that the bars will be centered on the midpoint of each bin by default. Compare the output when you specify 10, 100, and 1000 bins. For more control, you can provide the range over which you wish to bin the data, or you can specify the edges of each bin yourself.

For control over presentation or for further analysis, you can get the counts and bins used to generate the plot by unpacking the return value of **`plt.hist`**. (See Section 6.1.4 for details.)

```
counts, bin_edges, _ = plt.hist(data)
```

The function returns three objects, so three variables are required to unpack its output. If you are interested only in the first two objects, you can assign the third to Python's dummy variable, named underscore, as shown here.

If you do not wish to generate a plot and only want the histogram data, you can use **`np.histogram`** instead. It is the same function that **`plt.hist`** uses to generate data for plotting:

```
counts, bin_edges = np.histogram(data)
```

Note that this function only returns two objects.

Your Turn 6C

> Try it, and inspect the return values to make sure you understand how these functions generate a histogram. Note, in particular, the numbers of elements in the two arrays.

Once you've created the binned data, you can then plot it in another style or transform it prior to plotting. The **`plt.hist`** (or **`np.histogram`**) function has simply done the sorting and counting for you. For example, you can make a custom bar graph by using **plt.bar**. The arguments of this function are an array of bar positions, an array of bar heights, and a single width or an array of widths, which **plt.bar** uses to draw rectangles. All the arrays must be the same length, so you have to discard the last element of `bin_edges` returned by **plt.hist** or **np.histogram**, which is the right-hand edge of the last bin.

There are many ways to present data. The following code uses **plt.bar** to generate a colorful plot in which the width of a bar in the histogram is proportional to its height:

```
bin_size = bin_edges[1] - bin_edges[0]
new_widths = bin_size * counts / counts.max()
plt.bar(bin_edges[:-1], counts, width=new_widths, color=['r','g','b'])
```

6.3.2 Finer control

As mentioned earlier, Python offers a lot of control over how your data is displayed. Sometimes you will wish to specify the edges of the bins or the range of values that each bin collects, not just the number of bins. Both **plt.hist** and **np.histogram** can accommodate this: Instead of passing an integer number of bins to either function, supply an array whose entries are the edges of the desired bins. You need to provide one more edge than the number of bins. (Each bin begins where the previous one ended, but you need to specify where the first one starts.) Be aware that Python will ignore any data points that fall outside the range you provide.

For example, to divide a collection of random numbers into bins that each span an inverse power of 2, such as $0, \frac{1}{128}, \frac{1}{64}, \frac{1}{32}, \ldots, \frac{1}{2}, 1$, you can use **np.logspace** to generate the bins:

```
log2bins = np.logspace(-8, 0, num=9, base=2)
log2bins[0] = 0.0                  # Set first bin edge to zero instead of 1/256.
plt.hist(data, bins=log2bins)
```

If you bin the data by using **np.histogram** and then do some additional analysis, you can use **plt.bar** to display it as a histogram. The following code will do this:

```
bin_size = bin_edges[1] - bin_edges[0]
plt.bar(bin_edges[:-1], counts, width=bin_size, align='edge')
```

By default, the **plt.bar** function draws each bar centered on the corresponding value in the first argument. This behavior is generally not appropriate for a histogram. The preceding code fragment specifies a different behavior with the *align* keyword argument. (Try omitting it to see the difference.)

It is also possible to create histograms in higher dimensions, in which data are binned along two or more different axes. NumPy provides **np.histogram2d** to bin a collection of xy pairs. You can display the result as a three-dimensional bar chart by using **Axes3D.bar**, or as a grid of colored pixels (a heat map) by using **plt.imshow**. (Section 8.1.1, page 96 gives an example.) **np.histogramdd** extends binning to any number of dimensions.

6.4 CONTOUR PLOTS AND SURFACES

In earlier chapters, we saw several methods for plotting data sets with a single independent variable, such as a time series. Such data lend themselves to two-dimensional plots. However, models with two or more independent variables require higher dimensional plots.

A function of two variables, $h(x, y)$, can be interpreted as a surface whose height over each point (x, y) has been specified, just as Earth's topography is specified by the altitude at a given latitude and longitude. Two useful graphical representations of the height are contour plots and surface plots. A contour plot is a two-dimensional drawing in which contour lines or color are used to represent the height, as in a topographic map. A surface plot is a 3D perspective drawing of the surface itself.

6.4.1 Generating a grid of points

In order to make a plot of $h(x, y)$, you must specify the heights at a finite set of points. Typically you'll want those points to form a grid in the xy plane. Python gives a convenient way to construct such a grid: First, set up a 1D array of x values covering the desired range and an analogous 1D array of y values. Then, call the function **np.meshgrid** to create the grid. If the first array has N entries and the second has M entries, then **np.meshgrid** will return two $M \times N$ arrays. The arrays contain the x and y coordinates of each grid point, respectively. Try

```
x_vals = np.linspace(-3, 3, 21)
y_vals = np.linspace(0, 10, 11)
X, Y = np.meshgrid(x_vals, y_vals)
```

Inspect `X` and `Y` to make sure you understand the result: **np.meshgrid** returned a list containing two arrays, which we unpacked and assigned to `X` and `Y`. Each is an array with $11 \times 21 = 231$ entries giving the x and y coordinates of the grid points.[8] You can now evaluate a function on these arrays, perhaps by using vectorized math, to produce a third array called `Z`, and then create contour or surface plots using `(X,Y,Z)`. Sometimes it is useful to create the intermediate array of distances, `R=np.sqrt(X**2+Y**2)`, if the height only depends on the distance from the origin.

6.4.2 Contour plots

Suppose that we wish to visualize the function $z(x, y) = \cos x \sin y$. Creating a contour plot is simple using the arrays `X` and `Y` generated above:

```
Z = np.cos(X) * np.sin(Y)
plt.contour(X, Y, Z)
```

The number of contour lines is 10 by default. To change this, add a fourth argument to the function call: `plt.contour(X,Y,Z,20)`. Instead of an integer, you can also pass an array that contains the "heights" (z values) for which you want to see contours. You can control the appearance of contour lines by using keywords, and you can even label the contour lines with the following commands:

```
# contour.py
cs = plt.contour(X, Y, Z, 10, linewidths=3, colors='k')
plt.clabel(cs, fontsize=10)
```

The **plt.contour** function returns a `ContourSet` object, and the **plt.clabel** command knows how to add labels to the contours in this object.

The **plt.contour** function draws only contour lines. There is a related function called **plt.contourf** that will draw filled contours. You can change the set of colors used by Py-

[8] Python sets `X[i,j]=x_vals[j]` and `Y[i,j]=y_vals[i]`. You can change this convention with the keyword argument *indexing*=`'ij'`, which will set `X[i,j]=x_vals[i]` and `Y[i,j]=y_vals[j]`.

Plot via the command **plt.set_cmap**. If you enter the name of a nonexistent color map, Python will return an error that lists all of the available color maps.

6.4.3 Surface plots

Creating a surface plot is similar to creating a contour plot. The arguments to the functions are the same, but PyPlot has to access functions from a different module in order to create three-dimensional graphics. Recall Section 4.3.5:

```
from mpl_toolkits.mplot3d import Axes3D        # Import 3D plotting tool.
```

An **Axes3D** object can draw a variety of three-dimensional plots. For example, the following commands will create a surface plot:

```
ax = Axes3D( plt.figure() )          # Create 3D plotter attached to new figure.
ax.plot_surface(X, Y, Z)             # Generate 3D plot.
```

It takes a lot of computational resources to create a three-dimensional plot, so Python tries to use a small number of points when generating a surface. Its default is to use just one percent of the points you provide by skipping ten rows and ten columns at a time between points. This is good if you have a huge array, but if you have just 20 points along each direction, your surface will probably look terrible. To use all of your points, supply the keyword arguments *rstride* and *cstride* ("row stride" and "column stride") with the value 1:

```
ax.plot_surface(X, Y, Z, rstride=1, cstride=1)
```

Your Turn 6D

Make a surface plot of the function $z = x^2 + y^2$ over a suitable grid of values, where x and y range from -1 to 1.

The default of **plot_surface** is to use a single color and shading from imaginary sources of light to render a surface. You can get a more colorful plot by supplying a color map with the *cmap* keyword.[9] If you will be printing an assignment in grayscale, you may get better results if you use a colorless color map like `'gray'` or `'bone'`.

6.5 NUMERICAL SOLUTION OF NONLINEAR EQUATIONS

It is often necessary to solve nonlinear equations when studying physical and biological systems. For example, determining the fixed points in a single-gene toggle system requires finding the roots of a sixth-order polynomial. There are no general analytic solutions for polynomials of degree greater than four, and even for cubic equations the formulas are cumbersome. Biological oscillators may involve still more complicated functions that are not even polynomials. Numerical methods for finding the roots of such expressions are quite useful.

---

[9] In the blog that accompanies this book, we describe other options for coloring and shading 3D surfaces. (See press.princeton.edu/titles/11349.html.)

6.5.1 General real functions

The methods developed to find the roots of equations are closely related to methods for optimizing functions. Neither Python nor NumPy has a collection of optimization tools, but SciPy has an extensive library called `scipy.optimize`.[10] To gain access to the functions in this module, we need to import them. You can import the entire module:

```
from scipy import optimize
```

The command `dir(optimize)` will tell you the names of all of the functions available in this module, and `help(optimize)` will provide the details. However, we will not need most of the functions in this library, so we will simply import individual functions as they are required.

A useful function for finding roots is called `fsolve`. It takes two arguments: a function name and a point (or array of points) at which `fsolve` begins searching for roots.

You can define the function to be solved by using the approach described in Section 6.1. If you need to solve a relation like $g(x) = 7$, define your function as $f(x) = g(x) - 7$. The `fsolve` function determines the zeroes of whatever function you provide.

While `fsolve` is quite good at finding roots, it will do even better if you help it. You can provide an array of points *near* the zeros of a function, then have `fsolve` refine these. A good way to generate initial guesses is to plot the function and estimate visually where it crosses the x axis.

Example: Try the following and explain the results:

```
from scipy.optimize import fsolve
def f(x): return x**2 - 1
fsolve(f, 0.5)
fsolve(f, -0.5)
fsolve(f, [-0.5, 0.5])
```

Solution: The equation $x^2 - 1 = 0$ has two roots, but which root `fsolve` returns depends on where we tell the function to begin searching. Providing an array of starting points yields an array of roots found from those starting points.

Example: Now try the following, and explain what you see:

```
def f(x): return np.sin(x)**10
fsolve(np.sin, 1)
fsolve(f, 1)
```

Solution: The last expression above does not return exactly 0. What you are seeing is an example of *numerical error*, caused by the finite number of digits that Python uses to represent a number. The result is a number that is close the correct solution, but not exact.

The phrase "numerical error" is standard, but unfortunate. *You* didn't make any error; neither did Python. Your computer's hardware simply has limited precision.

A more serious type of error can occur when searching for the roots of a function with a singularity. Inspect the output of `fsolve(f,2)` when $f(x) = 1/(x - 1)$. This function has a discontinuity at $x = 1$ and never crosses the x axis. However, `fsolve` returns a solution without raising any errors or warnings. We can learn more by using a keyword to instruct `fsolve` to provide more information:

---

[10] T2 Its functions are adapted from the highly optimized MINPACK library of FORTRAN functions.

```
def f(x): return 1/(x-1)
fsolve(f, 2, full_output=True)
```

This time, the value of the root is returned along with an object that contains a lot more information. You can refer to `help(fsolve)` to learn what the terms mean, but the last line is a sure sign of trouble: The function reached its maximum number of evaluations without satisfying its requirements for a root of the equation. If your numerical result doesn't make sense, set `full_output=True` and consult `help(fsolve)`.

6.5.2 Complex roots of polynomials

The solutions returned by `fsolve` are restricted to be real numbers. Consider the equation $1/x = 1+x^3$. In order to use `fsolve`, we must manipulate this into $f(x) = x(1+x^3) - 1$ and solve for the roots. Assume that we have some good reason to guess that the real roots are near 1 and -1:

```
def f(x): return x * (1 + x**3) - 1
fsolve(f, 1)
fsolve(f, -1)
```

A quartic equation has four roots; but, in the present case, `fsolve` can only find two of them because the other two are complex.[11]

You can find all of the roots—real or complex—of any *polynomial* by using the NumPy function `np.roots`. This function takes a vector of coefficients that define a polynomial (see `help(np.roots)` for details). You can find all of the roots of the polynomial above with `np.roots([1,0,0,1,-1])`.

For arbitrary nonlinear equations (those that are not polynomials) with complex roots, Wolfram Alpha and *Mathematica* are reliable tools. For example, the function $\frac{1}{x} = 1 + x^{2.4}$ has three roots, but only one real root. It cannot be completely solved by `fsolve`.[12]

6.6 SOLVING SYSTEMS OF LINEAR EQUATIONS

Often we need to solve a set of simultaneous linear equations. For example, a least-squares fit of a two-parameter linear model to a data set involves solving equations of the general form

$$\begin{bmatrix} a_1 \\ a_2 \end{bmatrix} = \begin{bmatrix} c_{11} & c_{12} \\ c_{21} & c_{22} \end{bmatrix} \begin{bmatrix} x_1 \\ x_2 \end{bmatrix}.$$

We are given (or can find) the a's and c's, and wish to know the x's.

Python and NumPy do not provide a collection of linear algebra tools, but SciPy provides a library for matrix mathematics called `scipy.linalg`.[13] To gain access to the functions in this module, we need to import them. You can import the entire module.

```
from scipy import linalg
```

---

[11] One way to see this is to use a symbolic math program like Wolfram Alpha. See Section 8.4.

[12] T2 Actually, you *can* solve this particular equation by using `np.roots`. Rewrite it as $x^{3.4} = 1 - x$, raise both sides to the fifth power to generate a 17th-order polynomial, get all 17 roots from `np.roots`, then use a `for` loop to substitute them into the original expression and discard the spurious solutions.

[13] Its functions are adapted from the highly optimized LAPACK library of FORTRAN functions.

To learn more about the module and the functions it provides, use `dir(linalg)` and `help(linalg)`. Because we will only use a few of these functions, we will simply import them by name as they are needed. Some of the more commonly used are

`inv`	matrix inverse
`det`	determinant
`sqrtm`	matrix square root
`expm`	matrix exponentiation
`eig`	eigenvalues and eigenvectors of a matrix
`eigh`	eigenvalues and eigenvectors of a Hermitian matrix
`svd`	singular value decomposition

To solve the linear system above, we can compute the inverse of the matrix C and multiply the vector $\boldsymbol{a}$ by this matrix.

```
# matrix_inversion.py
from scipy.linalg import inv
a = np.array( [-1, 5] )
C = np.array( [ [1, 3], [3, 4] ] )
x = np.dot(inv(C), a)
```

This code puts the solution to the problem into the array `x`, as you can check by computing `np.dot(C,x)-a`. (Again, the result may not be exactly zero, due to numerical error. However, the difference should be on the order of 10^{-15} or less.)

The order of multiplication is important: The solution of $\boldsymbol{a} = \mathsf{C} \cdot \boldsymbol{x}$ is obtained by multiplying both sides *on the left* by C^{-1}.

6.7 NUMERICAL INTEGRATION

Integrating a function is a common task in physical modeling. For example, you may wish to integrate a probability density function to determine the probability that its variable will fall within a certain range of values.

As with linear algebra, Python and NumPy do not support numerical integration, but SciPy does. The appropriate module is called `scipy.integrate`.

```
from scipy import integrate
```

The command `dir(integrate)` will tell you the names of all of the functions in this module, and `help(integrate)` will provide more information. However, we will not need most of the functions in the library, so we will simply import individual functions as they are required.

There are several integration routines available in the module, each with particular strengths. The all-purpose workhorse is called `quad`, from "quadrature," an old word for integration.[14]

6.7.1 Integrating a predefined function

To carry out a numerical integration, you must provide a function name for the integrand and the limits of integration. Optional arguments allow more control over the integration. To see how this

[14] This function is adapted from the highly optimized QUADPACK library of FORTRAN functions.

works, let's evaluate

$$\int_0^{x_{\max}} dx \cos(x)$$

for various values of $x_{\max}$. Try this code:

```
# quadrature.py
from scipy.integrate import quad
x_max = np.linspace(0, 3*np.pi, 16)
integral = np.zeros(x_max.size)
for i in range(x_max.size):
    integral[i], error = quad(np.cos, 0, x_max[i])
plt.plot(x_max, integral)
```

Here is how the code operates:

- The second line imports the **quad** function.
- The third line creates an array of values for `x_max`.
- The fourth line sets up an array of the same shape as `x_max` to store the results.
- The next two lines carry out the integrations and store the results. By default, **quad** returns two values: the result of the integral and an estimate of the error. Line 6 unpacks these values. (The value of the error is overwritten during each iteration.)

The **quad** function evaluates **np.cos** at a carefully selected series of points and uses the result to approximate the integral over the range given by its second and third arguments.

Your Turn 6E

Do the integral analytically, and check whether **quad** got it right.

6.7.2 Integrating your own function

Next, suppose that you wish to integrate a function that is not predefined. If your function is short, you can define it at the command prompt:

```
def f(x): return x**2
```

If the function is long and complicated, or if you plan on using it more than once, you'll want to define it in a script or module, as described in Section 6.1. Once you have defined your function, you can pass its name (with no arguments or parentheses) to **quad** and integrate it like any other function.

Be aware that **quad** will only integrate real functions. To integrate a complex function, integrate its real and imaginary parts separately. If you give quad a function that returns complex values, it issues a warning but does not halt your program.

Your Turn 6F

a. Integrate $f(x) = x^2$ from 0 to 2, and check that **quad** gets it right.
b. Try a function whose integral you *don't* know: Evaluate

$$\int_0^{x_{\rm max}} \mathrm{d}x\, \mathrm{e}^{-x^2/2}$$

for values of $x_{\rm max}$ from 0 to 5, and plot the answers.

6.7.3 Oscillatory integrands

Sometimes we wish to evaluate an integral whose integrand oscillates rapidly. This can cause **quad** to fail to converge using its default settings. However, **quad** is an adaptive method, and if you allow it to subdivide the integration range sufficiently, it will generally yield satisfactory results. The keyword *limit* controls how fine a grid **quad** is allowed to use. Try

```
quad(np.cos, 0, 5000)              # Results in a warning and enormous error!
quad(np.cos, 0, 5000, limit=1000)  # No warning; accurate result
np.sin(5000)                       # Exact result for comparison
```

6.7.4 T2 Parameter dependence

The **quad** function can only integrate functions of a single variable. If you have a function of several variables and you wish to integrate one of them, holding the others fixed, there are two options: You can define a dummy function of one variable and pass the dummy function to **quad**, or you can supply a keyword and value to **quad** that will specify the constant arguments. For example, suppose that you have defined a function as follows:

```
def f(x, a, b, c): return a*x**2 + b*x + c
```

To integrate the function over x holding $(a, b, c) = (1, 2, 3)$, the two options work as follows:

```
# Use dummy function.
def g(x): return f(x, 1, 2, 3)
integral1, err = quad(g, -1, 1)
# Use keyword.
integral2, err = quad(f, -1, 1, args=(1, 2, 3))
```

If the variable you wish to integrate is not the *first* argument of the function, you cannot use the second method. You must define a dummy function.

6.8 NUMERICAL SOLUTION OF DIFFERENTIAL EQUATIONS

In the physical and life sciences, it is often possible to write down a system of differential equations that govern a system or describe the behavior of a model. However, it may be impossible to solve this system in terms of known functions. A classic example is the three-body problem in classical mechanics: $\boldsymbol{F} = m\boldsymbol{a}$ and Newton's law of gravitation are all that is required to write down the differential equations, but no one in the last four centuries has been able to find a general solution!

Numerical solution is a powerful tool in studying such systems.[15] Starting from some initial configuration of a system, you can calculate the next configuration from the given differential equations. From *that* configuration, you can calculate the *next* configuration by again using the equations, and so on. Computers are ideal for executing this repetitive operation, and efficient libraries have been developed for the task.

To solve differential equations, we turn once again to SciPy. The module we need is called **scipy.integrate**, the same module that contains **quad**. There are several functions available in the module, but we will restrict our attention to the one called **odeint**:[16]

```
from scipy.integrate import odeint
```

6.8.1 Reformulating the problem

An ordinary differential equation (ODE) is one that describes a function of a single independent variable, which we'll call $y(t)$. A textbook example is the driven harmonic oscillator:

$$\frac{\mathrm{d}^2 y}{\mathrm{d}t^2} = -y + g(t).$$

The **odeint** function can "only" solve ODEs of the form

$$\frac{\mathrm{d}\boldsymbol{y}}{\mathrm{d}t} = \boldsymbol{F}(\boldsymbol{y}, t), \tag{6.1}$$

where $\boldsymbol{y}$ is a vector whose components y_i are functions of t, and $\boldsymbol{F}$ is a vector whose components are functions of y_i and t. Fortunately, *any* explicit ODE can be put into this form. For example, to describe the second-order ODE for the driven harmonic oscillator to **odeint**, we must reformulate it as a system of coupled first-order equations. First, define two new variables that will be the components of the vector $\boldsymbol{y}$:

$$y_1 = y \qquad\qquad y_2 = \frac{\mathrm{d}y_1}{\mathrm{d}t}.$$

Next, write the derivatives of y_1 and y_2 in terms of y_1, y_2, and t:

$$\frac{\mathrm{d}y_1}{\mathrm{d}t} = \frac{\mathrm{d}y}{\mathrm{d}t} = y_2 \qquad\qquad \frac{\mathrm{d}y_2}{\mathrm{d}t} = \frac{\mathrm{d}^2 y_1}{\mathrm{d}t^2} = \frac{\mathrm{d}^2 y}{\mathrm{d}t^2} = -y + g = -y_1 + g.$$

This allows us to cast our problem, a second-order differential equation, in the form required by **odeint** (Equation 6.1): $\boldsymbol{y}$ is an array with two entries, and $\boldsymbol{F}(\boldsymbol{y}, t)$ is a function that returns an array with two entries:

$$\boldsymbol{y} = \begin{bmatrix} y_1 \\ y_2 \end{bmatrix} \qquad\qquad \frac{\mathrm{d}\boldsymbol{y}}{\mathrm{d}t} = \boldsymbol{F}(\boldsymbol{y}, t) = \begin{bmatrix} y_2 \\ -y_1 + g \end{bmatrix}. \tag{6.2}$$

Most common ODEs can be recast in a similar manner.

---

[15] Many authors use the phrase "numerical integration of an ODE" to denote the procedure we call "numerical solution of an ODE." We will reserve the word "integration" for actually performing integrals, a procedure that sometimes, but not always, can be used to solve ODEs.

[16] [T2] This function is adapted from the highly optimized ODEPACK library of FORTRAN functions. Calling **odeint** invokes a routine called LSODA, an adaptive solver that chooses a predictor-corrector method or a backward differentiation formula depending on its progress with the problem.

6.8.2 Solving an ODE

You must supply three arguments to **odeint**: the name of the function $\boldsymbol{F}(\boldsymbol{y}, t)$, an array $\boldsymbol{y}(t_0)$ that defines the initial conditions, and an array of t values at which you'd like **odeint** to evaluate the solution $\boldsymbol{y}(t)$. The general function call is

```
y = odeint(F, y0, t)
```

The variables in the expression are:

- **F** – A function $\boldsymbol{F}(\boldsymbol{y}, t)$ that accepts a 1D array and a scalar and returns an array.
- **y0** – A one-dimensional array with the initial values of $\boldsymbol{y}$.
- **t** – An array of t values at which $\boldsymbol{y}$ is to be computed. The first entry of this array is the time at which the initial values `y0` apply.
- **y** – An array of the values of $\boldsymbol{y}(t)$ at the points specified in `t`.

To determine the motion of an undriven harmonic oscillator (the example above with $g(t) = 0$), we first need to define the function $\boldsymbol{F}(\boldsymbol{y}, t)$. The first argument, `y`, is an array that contains the values of y_1 and y_2 at time t. The function we need is

```
# simple_oscillator.py
def F(y, t):
    """
    Return derivatives for second-order ODE y'' = -y.
    """
    dy = [0, 0]             # Create a list to store derivatives.
    dy[0] = y[1]            # Store first derivative of y(t).
    dy[1] = -y[0]           # Store second derivative of y(t).
    return dy
```

Note that **odeint** requires `F(y,t)` to accept time as an argument, even though the value of `t` is not used in this example. You will get an error if you omit this argument in the definition of `F(y,t)`.

We can now study the motion of the harmonic oscillator for different initial conditions:

```
# solve_ode.py
""" ODE solver for harmonic oscillator. """

import numpy as np, matplotlib.pyplot as plt
from scipy.integrate import odeint

# Import ODE to integrate:
from simple_oscillator import F

# Create array of time values to study:
t_min = 0; t_max = 10; dt = 0.1
t = np.arange(t_min, t_max+dt, dt)
```

```
# Provide two sets of initial conditions:
initial_conditions = [ (1.0, 0.0), (0.0, 1.0) ]

plt.figure()     # Create figure; add plots later.
for y0 in initial_conditions:
    y = odeint(F, y0, t)
    plt.plot(t, y[:, 0], linewidth=2)

skip = 5
t_test = t[::skip]                             # Compare at a subset of points.
plt.plot(t_test, np.cos(t_test), 'bo')         # Exact solution for y0 = (1,0)
plt.plot(t_test, np.sin(t_test), 'go')         # Exact solution for y0 = (0,1)
```

This code illustrates the following:

- Like **quad**, **odeint** expects its first argument to be a function name (with no parentheses).
- The second argument is a vector with two entries, specifying the initial conditions.
- The return value of **odeint** is an array. The first column is $y(t)$ evaluated at the values in `t`; the second is $\mathrm{d}y/\mathrm{d}t$. If you solve a higher order ODE, the third column will contain $\mathrm{d}^2y/\mathrm{d}t^2$, and so on.

Your Turn 6G

> Modify `simple_oscillator.py` to incorporate the driving force $g(t) = \sin(0.8t)$, and comment on how the solution changes.

6.8.3 T2 Parameter dependence

If we wanted to explore several different harmonic oscillators, we might consider modifying the function `F(y,t)` as follows:

```
# parametric_oscillator.py
def F(y, t, spring_constant=1.0, mass=1.0):
    """
    Return derivatives for harmonic oscillator:
        y'' = -(k/m) * y
    y = displacement in [m]
    k = spring_constant in [N/m]
    m = mass in [kg]
    """
    dy = [0, 0]           # Array to store derivatives
    dy[0] = y[1]
    dy[1] = -(spring_constant / mass) * y[0]
    return dy
```

To specify fixed parameters when using **odeint**, there are two options, just as for **quad** in Section 6.7.4. First, we can define a dummy function that sets the parameters and pass this function to **odeint**. Alternatively, we can supply an optional argument to **odeint** by using the keyword *args*. Suppose that the system has a spring constant of 2.0 N/m and a mass of 0.5 kg.

```
y0 = (1.0, 0.0)
t = np.linspace(0, 10, 101)

# Use dummy function.
def G(y, t): return F(y, t, 2.0, 0.5)
yA = odeint(G, y0, t)

# Use keywords.
yB = odeint(F, y0, t, args=(2.0, 0.5))
```

To use the second method, the parameters you wish to set must come after `y` and `t` in the list of arguments to `F`. Otherwise, you must define a dummy function.

6.9 VECTOR FIELDS AND STREAMLINES

A vector field is a function whose value at any point in space is a vector. Common examples from physics include the electric field, the gravitational field, and the velocity field of a fluid. PyPlot provides two useful functions for visualizing vector fields and their streamlines: **plt.quiver** and **plt.streamplot**.

6.9.1 Vector fields

You can plot a two-dimensional vector field with **plt.quiver**. This function requires four arguments, all of which are arrays of the same size. The first two arguments define a grid of (x, y) values. (See Section 6.4.3.) Instead of specifying an ordinary function like height or temperature at these points, the next two arguments specify the components of a two-dimensional *vector* $\boldsymbol{v}$ at the corresponding (x, y) coordinates. Python will draw an arrow starting at each grid point to represent the vector field $\boldsymbol{v}(x, y)$.

Your Turn 6H

Try the following:

```
# vortex.py
coords = np.linspace(-1, 1, 11)
X, Y = np.meshgrid(coords, coords)
Vx, Vy = Y, -X
plt.quiver(X, Y, Vx, Vy)
```

Explain the "vortex" pattern you get.

Python's default arrows may not be what you need. **help**(**plt.quiver**) reveals several options for controlling the appearance of the arrows. For example, if you want your arrows to be the length that you specified in your arrays, the keyword arguments *units*='xy' and *scale*=1 will produce this behavior. If the components of the vector are (1,0) at some point, the corresponding arrow will have a length of 1.0 relative to the scale used to draw the axes. (This means the arrows will get larger as you zoom in and smaller as you zoom out.)

Vector fields often arise as the gradient of a scalar function. For example, Fick's law states that the flux of particles at a point is proportional to the gradient of the concentration:

$$\boldsymbol{J} = -D\boldsymbol{\nabla} c.$$

NumPy can estimate the gradient of a function evaluated on a grid via **np.gradient**.[17] The following code displays the gradient of a bell curve on its contour plot. Consult **help**(**np.gradient**) to explore its arguments and return value.

```
# gradient.py
import numpy as np, matplotlib.pyplot as plt

coords = np.linspace(-2, 2, 101)
X, Y = np.meshgrid(coords[::5], coords[::5])     # Coarse grid for vector field
R = np.sqrt(X**2 + Y**2)
Z = np.exp(-R**2)
x, y = np.meshgrid(coords, coords)               # Fine grid for contour plot
r = np.sqrt(x**2 + y**2)
z = np.exp(-r**2)

ds = coords[5] - coords[0]                       # Coarse grid spacing
dX, dY = np.gradient(Z, ds)                      # Calculate gradient.

plt.contourf(x, y, z, 25)
plt.set_cmap('coolwarm')
plt.quiver(X, Y, dX.transpose(), dY.transpose(), scale=25, color='k')
```

Because of the way NumPy calculates the gradient and our convention for x and y axes, we must take the *transpose* of the arrays that **np.gradient** returns—that is, we must interchange rows and columns.

6.9.2 Streamlines

A system of first-order ordinary differential equations defines a vector field. One way to find solutions is to follow the arrows, generating "streamlines." The word comes from an analogy to water flow: The velocity of the water defines a vector field; the trajectory of any droplet of water is a streamline. Electric and magnetic "field lines" are also streamlines.

Python's **plt.streamplot** function will generate several trajectories from a vector field. The syntax is

```
plt.streamplot(x, y, Vx, Vy)
```

You can adjust the length and density of lines by using optional arguments. See **help**(**plt.streamplot**) for details. The four arguments are the same as those used in **plt.quiver**. They specify the coordinate grid and the vector field to follow.[18]

To illustrate the use of this function, begin with the vector field you drew in Section 6.9.1:

[17] An analytic formula for the gradient of a function is usually more accurate than **np.gradient**, but such formulas are not available for experimental measurements.

[18] The documentation for **plt.streamplot** specifies 1D arrays for x and y; however, passing 2D arrays generated by **meshgrid** works just as well.

```
# streamlines.py
lower, upper, step = -2, 2, 0.1
coords = np.arange(lower, upper + step, step)
X, Y = np.meshgrid(coords, coords)
Vx, Vy = Y, -X
plt.streamplot(X, Y, Vx, Vy, linewidth=2)
```

Notice that we chose a finer grid of points than before. Such a fine grid would have made a cluttered picture of the vector field with too many arrows. For generating streamlines, a fine grid yields more accurate results. (Python doesn't need to interpolate as much as it would with a coarse grid.)

Your Turn 6I

a. Perhaps the picture drawn in the preceding example was a bit too predictable. Replace line 5 with

```
Vx = Y - 0.1 * X
Vy = -X - 0.1 * Y
```

and explain the streamlines that you find.
b. Next, replace line 5 with

```
Vx, Vy  = X, -Y
```

and explain what happens.

CHAPTER 7

Second Computer Lab

In this lab, you'll use Python to generate two-dimensional random walks, plot their trajectories, and look at the distribution of end points for a large number of random walkers. Our goals are to

- Generate random walk trajectories that begin at the origin and take random diagonal steps:

$$x_{n+1} = x_n \pm 1 \qquad y_{n+1} = y_n \pm 1. \tag{7.1}$$

- Plot trajectories of four such walks in separate subplots of a single figure.
- Plot the end points of many trajectories in a single figure to see how they are distributed.
- Compute the average final distance of the walkers from the origin.

First, review Section 6.2.

7.1 GENERATING AND PLOTTING TRAJECTORIES

Our first task is to create a random walk of 1000 steps, each given by Equation 7.1. Each trajectory will be a list of 1000 x values and 1000 y values. It's good programming practice to define the size of the simulation at the beginning of a script:

```
num_steps = 1000
```

Now you can set the size of all arrays by using `num_steps`.

Assignment:

a. *Use the ideas in Section 6.2 to make a random walk trajectory, and then plot it.* To remove any distortion, use the command `plt.axis('equal')` after making the plot.

b. *Now make four such trajectories, and look at all four side by side.* Use `plt.figure()` to create a new figure window. You can access the individual subplots by using commands like `plt.subplot(2,2,1)` before the first `plt.plot` command, `plt.subplot(2,2,2)` before the second `plt.plot` command, and so on. Python may give each plot a different magnification. Consult `help(plt.xlim)` and `help(plt.axis)` to find out how to give each of your plots the same x and y limits.

7.2 PLOTTING THE DISPLACEMENT DISTRIBUTION

Run your script several times, and compare the resulting trajectories. Your four plots always look different! Sometimes the walker wanders off the screen; sometimes it remains near the origin. And yet, there is some family resemblance between them. Let us begin to understand in what sense they resemble each other.

How far does the random walker get after 1000 steps? More precisely, what is the distance from the starting point $(0, 0)$ to the ending point (x_{1000}, y_{1000}), for each of many random walks? Instead of four walks, we now want many, say, 100. You could manually examine all 100 plots, but it would be hard to see the common features. Instead, we'll ask Python to generate all of these random walks, but show us only a summary.

One straightforward way to make 100 walks is to take the code you already wrote and embed it inside of a **for** loop. You could create three arrays, `x_final`, `y_final`, and `displacement`, to store the ending x and y positions and distance to the origin. Then, just before the end of the loop, you could add something like

```
x_final[i] = x[-1]
y_final[i] = y[-1]
displacement[i] = np.sqrt(x[-1]**2 + y[-1]**2)
```

(You can also solve the problem by using vectorized operations instead of a **for** loop. Try to figure out how. The vectorized approach is faster than a loop if the arrays are not too large—that is, if `num_walks*num_steps` is smaller than 10^7.)

We can summarize the results in at least three ways. You have a lot of end points (`x_final`, `y_final` pairs), so you can make a *scatter plot* by using **`plt.plot`** or **`plt.scatter`**. Alternatively, you can examine the lengths of the final displacement vectors, or their squares.

Assignment:

a. *Once you have a code that works, increase the number of random walks from 100 to 1000. (See Section 3.3.5.) Make a scatter plot of the end points.*

b. *Use **`plt.hist`** to make a histogram of the `displacement` values.*

c. *Make a histogram of the quantity* `displacement**2`.

d. *Your result from (c) may inspire a guess as to the mathematical form of the histogram. Try semilog and log-log axes to inspire and test your guess.*

e. *Use **`np.mean`** to find the average value of* `displacement**2` *(the* **mean-square displacement***) for a random walk of 1000 steps.*

f. *Find the mean-square displacement of a 4000-step walk.* If you wish to carry the analysis further, see if you can determine how the mean-square displacement depends on the number of steps in a random walk.

It turns out that random walks are partially predictable after all. Out of all the randomness comes systematic *statistical* behavior, partly visible in your answers to (b–f).

Experimental data also agree with these predictions. The random walk, although stripped of much of the complexity of real Brownian motion, nevertheless captures nontrivial aspects of Nature that are not self-evident from its formulation. See if your output qualitatively resembles the experimental data shown in Figure 7.1 for the diffusion of a micrometer-size particle in water.

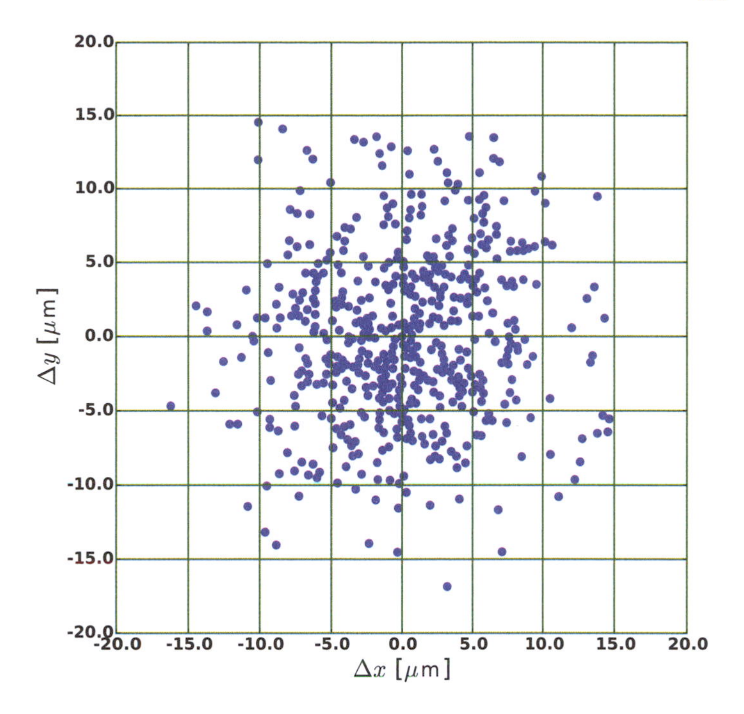

Figure 7.1: Experimental data for Brownian motion. Each dot is the final position after a fixed amount of time for a particle that started at the center of the figure.

7.3 RARE EVENTS

7.3.1 The Poisson distribution

Imagine an extremely unfair coin that lands heads with probability ξ equal to 0.08 (not 0.5). Each trial consists of flipping the coin 100 times. You might expect that we'd then get "about 8" heads in each trial, although we could, in principle, get as few as 0, or as many as 100.

The **Poisson distribution** is a discrete probability distribution that applies to rare events.[1] For our extremely unfair coin, the Poisson distribution predicts that the probability of the coin coming up heads ℓ times in 100 flips is

$$\mathcal{P}(\ell) = \frac{\mathrm{e}^{-8} \cdot 8^{\ell}}{\ell!}, \tag{7.2}$$

where ℓ is an integer greater than or equal to 0.

[1] Nelson, 2015, Chapter 4 discusses this distribution.

Assignment:

a. *Before you start flipping coins, plot this function for some interesting range of ℓ values.* You may find the following helpful:
   - The `factorial` function can be imported from `scipy.special`.
   - You need not take values of ℓ all the way out to infinity. You'll see that $\mathcal{P}(\ell)$ quickly gets negligibly small.
   - In Python, the elements of a vector are always numbered 0, 1, 2, 3,. . . , and ℓ is also an integer starting from zero, so ℓ is a good array index.
   - The value of 8^ℓ can get *very* large—larger than the largest integer NumPy can store.[2] To avoid numerical *overflow*—and erroneous results—use an array of floats instead of integers. Consult `help(np.arange)`, and read about the *dtype* keyword argument.

b. *Perform N coin flip trials, each consisting of 100 flips of a coin that lands heads only 8% of the time.* [Good practice: Eventually you may want to take N to be a huge number. While developing your code, make it not so huge, say, $N = 1000$, so that your code will run fast.]

c. *Get Python to count the number of heads, M, for each trial. Then, use* `plt.hist` *to create a histogram of the frequency of getting M heads in N trials.* If you don't like what you see, consult `help(plt.hist)`. (For example, `plt.hist` may make a poor choice about how to bin the data.)

d. *Make a graph of the Poisson distribution (Equation 7.2 above) multiplied by N.* What's the most probable outcome? Graph this plot on the same axes as the histogram in (c).

e. *Repeat (b–d) for $N = 1\,000\,000$, and comment.* (This may take a while.)

Click the RUN▶ button over and over for $N = 1000$, and observe that the distribution is a bit different every time, and yet each plot has a general similarity to the others.

7.3.2 Waiting times

If we flip our imagined coin once every second, then our string of heads and tails becomes a time series called a **Poisson process**, or *shot noise*. Flipping heads is a rare event, because $\xi = 0.08$. We expect long strings of tails, punctuated by occasional heads. This raises an interesting question: After we get a heads, how many flips go by before we get the next heads? More precisely, what's the *distribution* of the *waiting times* from one heads to the next?

Here's one way to get Python to answer that question. We can make a long list of ones and zeros, then search it for each occurrence of a 1 by using NumPy's `np.nonzero` function. This function takes an array of numbers and returns an array of the indices of its nonzero entries. Experiment with small arrays like `np.nonzero([1,0,0,-1])` to understand its behavior. Consult `help(np.nonzero)` for more information.

Each waiting time is the length of a run of zeros, plus one. You can subtract successive entries in the array returned by `np.nonzero` to find the waiting time between successive heads, then make a graph showing the frequencies of those intervals. NumPy's `np.diff` function will take the difference of successive entries in an array. Flatten the array returned by `np.diff` before plotting. See Section 2.2.8 for details on flattening arrays. Consult `help(np.diff)` for more information about that function.

---

[2] NumPy uses 64-bit integers, so the largest number it can store is $2^{63} - 1$.

Try to guess what this distribution will look like before you compute the answer. Someone might reason as follows: "Because heads is a rare outcome, once we get a tails we're likely to get a lot of them in a row, so short strings of zeros will be less probable than medium-long strings. But eventually we're bound to get a heads, so *very* long strings of zeros are also less common than medium-long strings." Think about it. Is this sound reasoning? Now, compute your answer. If your output is not what you expected, try to figure out why.

Assignment:

a. *Construct a random string of 1's and 0's representing 1000 flips of the unfair coin. Then, plot the frequencies of waiting times of length 0, 1, 2, ..., as outlined above. Also make a semilog plot of these frequencies. Is this a familiar-looking function?*

b. *What is the average waiting time between heads?*

c. *Repeat (a) and (b) for* 1 000 000 *flips of the coin.*

CHAPTER 8

Still More Techniques

It's nice to know that the computer understands the problem.
But I would like to understand it, too.
—Eugene Wigner

This short chapter attempts to round out your Python toolkit. We introduce some tools for image processing and animation, and then discuss symbolic calculations.

8.1 IMAGE PROCESSING

A digital image is a collection of pixels. The image can be stored in a variety of formats. In the case of a black-and-white photograph, each pixel is represented by a number corresponding to the intensity (mean photon rate) sensed at a point in the xy plane of the detector.[1] A digital camera takes the light intensity in each pixel and reports it as an integer between 0 and $2^m - 1$, where m is called the *bit depth.* A common choice is $m = 8$ (256 distinct light levels). A color image consists of *three* such arrays, reporting the intensities of red, blue, and green light.

A computer uses the pixel data in an image file to reproduce the image on your monitor. As far as Python is concerned, then, *every image is an array of numbers.* Conversely, *any array with the appropriate shape and data content can be displayed as an image.* This mapping between arrays and images allows us to use Python to import, analyze, transform, and save images.

8.1.1 Images as NumPy arrays

Let's use Python to import and display an image file. We first need to give Python access to some image data. Follow the instructions of Section 4.1 (page 48) to obtain the data set `16catphoto`. Copy the file `bwCat.tif` into your working directory.

Neither basic Python nor NumPy contains modules for working with images, but PyPlot contains functions for reading, displaying, and saving image files: **`plt.imread`**, **`plt.imshow`**, and **`plt.imsave`**. Import the photograph to an array with the following command:

```
photo = plt.imread('bwCat.tif')
```

(If you have trouble loading image files, you may need to install the `pillow` library. See Section A.1.2 for details.) Verify that the photograph is indeed represented as an array of numbers by looking at some elements of `photo`. You should see the array in the Variable Explorer. You can also inspect the attributes of the array or examine a slice:

[1] We are discussing raster images, also called bitmaps, which are saved in formats like `.tif` or `.jpg`. Another category, called vector graphics, represents figures mathematically and is saved in formats like `.svg` or `.eps`; see Section 4.3.8.

```
photo.shape
photo.dtype
photo[:10, :10]
```

This photograph is represented as a 648×864 array of integers. The data type of the array is `uint8`, which means that each value is an unsigned integer represented by 8 bits, matching the bit depth of the original image.

8.1.2 Saving and displaying images

We can now use PyPlot to display the array as an image:

```
plt.imshow(photo)
```

The resulting figure is probably not what you expected. PyPlot's defaults are convenient for plotting mathematical functions, but not black and white photographs. The following commands will set the color map, remove the axes, and change the background color for more convenient viewing.

```
plt.set_cmap('gray')              # Use grayscale for black and white image.
plt.axis('off')                   # Get rid of axes and tick marks.
fig = plt.gcf()                   # Get current figure object.
fig.set_facecolor('white')        # Set background color to white.
```

Now we see a recognizable image of a cat. You may find it convenient to define a single function to perform all these steps and display a black and white image from an array.

You can export any figure in a photographic format by using the SAVE button in the figure window. You can also create an image file from the command prompt or within a script:

```
plt.imsave('cat.jpg', photo, cmap='gray')
```

The function **plt.imsave** treats the array in the same way as **plt.imshow**, and it will use the current color map. Thus, we need to specify an appropriate color map for a black and white image.

8.1.3 Manipulating images

Because Python converts an image into a numerical array, you can perform mathematical operations on the array, potentially enhancing the image.

Your Turn 8A

Try this:

```
new_cat = (photo < photo.mean())
```

Display the new array, and explain what you see. Compare `new_cat` and `photo` in the Variable Explorer.

In Chapter 9, you'll learn about some common techniques for manipulating and enhancing images. However, you should be aware of the limits imposed by scientific ethics on modifying images to be used as evidence. (See Cromey, 2010.)

8.2 DISPLAYING DATA AS AN IMAGE

If you generate a two-dimensional data set—a function of (x, y) coordinates, for instance—you may wish to display it as an image. You can do this using **plt.imshow**, but you have to provide extra instructions so that your data is displayed appropriately.

There is a significant difference between the conventions we normally use in mathematics to label coordinates and the way Python displays images. This can be very confusing if you are not aware of it. In mathematics, we usually place the origin, (0,0), at the lower left corner of a plot. Increasing the x-coordinate moves a point to the right, and increasing the y-coordinate moves a point up. Python's conventions for displaying an image are almost the exact opposite!

The element in `photo[0,0]` is displayed at the upper-left corner of the picture. Increasing the first index moves *down* the image, and increasing the second index moves *right* across the image. This seems strange until you recall that Python is not graphing a function; it is displaying an array. `photo[0,0]` is the first element of the array. The first index refers to a row in the array, and the second index refers to a column. So Python's convention is exactly what we would use if we were writing out the elements of a matrix.

Thus, if we wish to display data whose indices correspond to x, y values, we must first *transpose* the rows and columns of our data array. (Since the first index of an array controls the vertical position of an entry and the second controls the horizontal location, we need to switch the roles of x and y prior to using **plt.imshow**.) Second, we need to tell Python to put the origin of our coordinate system in the lower-left corner of the image. The keyword argument `origin='lower'` instructs Python to do this. After that, our conventions for labeling points will match up perfectly with Python's conventions for creating an image from array data.

The following example illustrates these ideas.

```
# heatmap.py
import numpy as np, matplotlib.pyplot as plt
plt.figure()
x = np.random.randn(5000) - 1                          # Generate random coordinates.
y = 2 * np.random.randn(5000)
# Create 2D histogram with same number of bins and same range for x and y.
counts, x_bins, y_bins = np.histogram2d( x, y, \
                              bins=[100,100], range=[(-5,5), (-5,5)])

# Display data as a heatmap. Provide legend for colors.
plt.imshow(counts.transpose(), origin='lower', cmap='hot')
plt.colorbar()
```

This code generates two sets of random coordinates drawn from the standard normal distribution using **np.random.randn**. (Subtracting 1 from the first array shifts the mean value of the x distribution to the left. Multiplying the second array by 2 makes the standard deviation of the y coordinates twice as large as that of the x coordinates. Thus, if the image is displayed properly, it should be shifted to the left and taller than it is wide.) Next, the code instructs Python to bin these data in a 2D histogram. Finally, the code calls **plt.imshow** to display the data as an image.

Explore different ways of displaying images and data sets to better understand Python's conventions. Try displaying the cat image as (x, y) data, or the raw (x, y) data as an image and interpret what you see.

8.3 ANIMATION

A picture may be worth a thousand words, but a motion picture can be even better. Matplotlib contains a module called **animation** for creating movies from plots. We can also view a collection of still plots (*frames*) in sequence to create a simple animation.

We provide two scripts to create animations below. They illustrate two different approaches, but they have a common design principle: Create an empty plot, take control of the line or point objects you wish to animate, then update them in each frame. Once you understand how to do this, you can create a variety of animations.

8.3.1 Creating animations

The following script will create a movie of one of the random walks you studied in Chapter 7 by using the function **FuncAnimation** from the **animation** module. Use Python's **help** function to explore the many options available.

```
# walker.py
# Jesse M. Kinder -- 2017
""" Make a movie out of the steps of a two-dimensional random walk. """
import numpy as np, matplotlib.pyplot as plt
from matplotlib import animation
from numpy.random import random as rand

# Set number of steps for each random walk.
num_steps = 100

# Create an empty figure of the desired size.
plt.close('all')          # Clear anything left over from prior runs.
bound = 20
fig = plt.figure()        # Must have figure object for movie.
ax = plt.axes(xlim=(-bound, bound), ylim=(-bound, bound))

# Create empty line and point objects with no data.
# They will be updated during each frame of the animation.
(my_line,) = ax.plot([], [], lw=2)                   # Line to show path
(my_point,) = ax.plot([], [], 'ro', ms=9)            # Dot to show current position

# Generate the random walk data.
x_steps = 2*(rand(num_steps) < 0.5) - 1         # Generate random steps +/- 1.
y_steps = 2*(rand(num_steps) < 0.5) - 1
x_coordinate = x_steps.cumsum()                 # Sum steps to get position.
y_coordinate = y_steps.cumsum()

# This function will generate each frame of the animation.
# It adds all of the data through frame n to a line
# and moves a point to the nth position of the walk.
def get_step(n, x, y, this_line, this_point):
    this_line.set_data(x[:n+1], y[:n+1])
    this_point.set_data(x[n], y[n])
```

```
35 # Call the animator and create the movie.
   my_movie = animation.FuncAnimation(fig, get_step, frames=num_steps, \
                       fargs=(x_coordinate, y_coordinate, my_line, my_point) )

   # Save the movie in the current directory.
40 # *** NEXT LINE WILL CAUSE AN ERROR UNLESS FFMPEG OR MENCODER IS INSTALLED. ***
   # my_movie.save('random_walk.mp4', fps=30)
```

This script uses a different approach to plotting to make sure that the fixed elements of the plot—axes, tick marks, legends, labels, and so on—are not changed from one frame to the next. Instead of drawing each frame "from scratch," the script creates a figure and axes just once (lines 14–15). Then, it creates two variables and assigns them to line and point objects that initially contain no data (lines 19–20).[2] The function `get_step` modifies the data of the line and point objects—an intentional side effect—but has no effect on the rest of the figure. The first argument of `get_step` is the frame number.

The name of the figure and the function for updating the plot are passed as arguments to **`animation.FuncAnimation`**, which calls `get_step` repeatedly to update the graph and generate the frames of the movie. We also provided additional arguments to specify the data, line, and point objects to use in the function with the keyword *`fargs`*. (The current frame number is automatically passed as the first argument. If the function requires additional arguments, as in this example, they should be passed by using *`fargs`*.)

8.3.2 Saving animations

To create and view an animation you don't need any software other than Matplotlib. You can share your code with others, but they will need Python to view your beautiful work. Sometimes it is more useful to save your animation in a format that anyone can view, such as a web page or a movie file. We will describe two options. The first can be implemented entirely in Python. The second requires an additional piece of software called an encoder.

HTML movies

A simple approach to making an animation is to create a flip book—a series of images that create the illusion of motion when viewed in rapid succession. You may have animated stick figures in this way when you were young, or bored.

You already know how to create a series of still images using Python. You just need a way to tie them all together and display them. Via `press.princeton.edu/titles/11349.html`, you can download a file called `html_movie.py`. This module contains a function called **`movie`** that creates a "web page" that will display a series of images in a continuous loop. You do not need to be online to view the animation—you just need a web browser such as Firefox, Safari, or Chrome.

The following script will create an HTML movie of two traveling waves that pass by each other.

```
# waves.py
# Jesse M. Kinder --- 2017
""" Generate frames for an animation of moving Gaussian waves. """
```

[2] We encountered a similar technique in Section 4.3.2 (page 59). The keywords *`lw`* and *`ms`* are abbreviations for *`linewidth`* and *`markersize`*, respectively.

```
import numpy as np
5 import matplotlib.pyplot as plt
# First obtain html_movie.py via http://press.princeton.edu/titles/11349.html
# or http://physicalmodelingwithpython.blogspot.com/ .
from html_movie import movie

10 # Generate waves for each frame.
# Return a Gaussian with specified center and spread using array s.
def gaussian(s, center=0.0, spread=1.0):
    return np.exp(-2 * (s - center)**2 / spread**2)

15 # All lengths are in [m], all times are in [s], and all speeds are in [m/s].
# Define the range of values to display.
x_min, x_max  = -4.0, 4.0
y_min, y_max  = -3.0, 3.0
# Define array of positions.
20 dx = 0.01
x = np.arange(x_min, x_max + dx, dx)

# Define the duration and number of frames for the simulation.
tmin, tmax  = 0.0, 4.0
25 num_frames = 100
t = np.linspace(tmin, tmax, num_frames)

# Define the initial position and speed of Gaussian waves.
r_speed = 2.0         # Speed of right-moving wave
30 r_0 = -4.0         # Initial position of right-moving wave
l_speed = -2.0        # Speed of left-moving wave
l_0 = 4.0             # Initial position of left-moving wave

# Generate a figure and get access to its Axes object.
35 plt.close('all')
fig = plt.figure(figsize=(10, 10))
ax = plt.axes(xlim=(x_min, x_max), ylim=(y_min, y_max))

# Create three empty line objects and grab control.
40 # The loop below will update the lines in each frame.
ax.plot([], [], 'b--', lw=1)          # Line for right-moving wave
ax.plot([], [], 'r--', lw=1)          # Line for left-moving wave
ax.plot([], [], 'g-', lw=3)           # Line for sum of waves
lines = ax.get_lines()                # Get list of 3 line objects in plot.
45
# It is essential that the frames be named in alphabetical order.
# {:03d} formats integers with three digits, including leading zeros if needed:
# '000_movie.jpg', '001_movie.jpg', and so on.
file_name = "{:03d}_movie.jpg"
50
# Generate frames and save each figure as a separate .jpg file.
for i in range(num_frames):
```

```
        r_now = r_0 + r_speed * t[i]                    # Update centers of waves.
        l_now = l_0 + l_speed * t[i]
55      yR =  gaussian(x, r_now)                         # Get current data for waves.
        yL = -gaussian(x, l_now)
        lines[0].set_data(x, yR)                         # Update right-moving wave.
        lines[1].set_data(x, yL)                         # Update left-moving wave.
        lines[2].set_data(x, yR + yL)                    # Update sum of waves.
60      plt.savefig(file_name.format(i))                 # Save current plot.

    # Use HTML movie encoder adapted from scitools to create an HTML document that
    # will display the frames as a movie.  Open movie.html in web browser to view.
    movie(input_files='*.jpg', output_file='movie.html')
```

To view the animation, open `movie.html` in a web browser. (To open a file from your computer in your web browser, choose the menu option `File>Open file ...`, and direct your browser to the file you have created.) This file must be located in the same folder as the image files to run properly. (It is not a stand-alone movie. It just tells your browser which image files to display and when to display them.) To share the animation, send the entire folder to a friend, or upload it to a web server.

In the last line, the keyword and value `input_files='*.jpg'` uses the asterisk as a wildcard to instruct the function to use *all* files in the current folder that have the `.jpg` extension. It will insert the files into the movie in alphabetical order. Line 49 formats the file names so that alphabetical and chronological order are the same. (If we had used the `{:d}` format specifier, there would be no leading zeros, and `10_movie.jpg` would come before `2_movie.jpg` in alphabetical order.)

The `html_movie` module was adapted from the `scitools` library developed by Hans Petter Langtangen.[3]

T2 *Using an encoder*

The approach in the preceding section generates an animation that you can share, but it doesn't create a standalone file in a standard video format suitable for embedding in another web page, uploading to YouTube, or editing with video software. (Perhaps you could add a soundtrack ...) To save the animation in a standard video format, you must extend Python's capabilities with a video **encoder**. FFmpeg, which can be downloaded at `www.ffmpeg.org`, is one Python-friendly option.

The last line of `walker.py` above, when uncommented, calls **`movie.save`**, which in turn invokes FFmpeg.

To use an encoder, you need to download and install it on your system, then make certain that Python can locate it. Using a package manager is the most straightforward way to do this. A package manager will determine all the libraries that are necessary to run the program you request, then download, install, and link everything together so your program runs properly. Section A.4 in Appendix A describes how to install FFmpeg.

Once you have successfully installed FFmpeg, you can uncomment the last line of `walker.py` above. Now when you run the script, Python will create a movie file called `random_walk.mp4` that you can edit, upload, and view in a movie player or web browser.

If you have difficulty linking Python to FFmpeg, you can run FFmpeg directly from your operating system's command line (not the IPython console). Make a series of frames with sequential names as in `waves.py` above. Then, use the following command to create the movie:

---

[3] `scitools` is a useful library, but as of this writing it is not compatible with Python 3. The entire library can be downloaded at `github.com/hplgit/scitools`. See also Langtangen, 2016.

```
$ ffmpeg -r 24 -i %03d_movie.jpg movie.mp4
```

The option `-r 24` specifies the frame rate in frames per second, and the `-i` option identifies the file names of the frames created by your code.

ImageMagick is another free software package that can be used to create animations at the command line. The software can be downloaded at `imagemagick.org`, or installed by using the `conda` package manager as described in Section A.4 (page 118). If you have installed ImageMagick, you can create a `gif` animation from the output of `waves.py` by using the following command:

```
$ convert -delay 1x24 *.jpg movie.gif
```

The `-delay 1x24` option instructs ImageMagick to wait 1/24th of a second between each frame, and `*.jpg` tells ImageMagick to use all file names—in alphabetical order—that have a `.jpg` extension. You can view `movie.gif` in any web browser, or you can embed it in a web page.

8.4 ANALYTIC CALCULATIONS

Computers were originally created for numerical calculations, but they can also do symbolic mathematics. In Python, we must import special libraries to do symbolic calculations. We can also take advantage of online resources. In this section, we describe both approaches.

8.4.1 The SymPy library

The SymPy library allows Python to do symbolic math.[4] To get some idea of what is possible with SymPy, try the following commands in the IPython console:

```
# sympy_examples.py
from sympy import *
init_session()
expand( (x + y)**5 )
factor( x**6 - 1 )
pi.n(100)
plot( besselj(0, x), besselj(1, x), (x, 0, 10) )
diff( x*sin(y), x, y )
integrate( cos(x)**2, x )
integrate( exp(-x**2), (x, -oo, oo) )
Sum( k**3, (k, 0, m) ).doit().factor()
dsolve( f(x).diff(x) + f(x) - x, f(x) ).simplify()
limit( sin(x)/x, x, 0 )
```

SymPy is a library where `import *` is so incredibly convenient that we make an exception to our general advice and bring all its functions into our working environment. When Python executes `init_session()`, it defines several variables and functions. The examples will not work if you omit this command. For more details, see `docs.sympy.org/latest/`.

SymPy is useful for quick symbolic calculations and plots of analytic functions. It is simple to use, and it is native to Python. Its functions can be imported and used in any script, and its output is available without copying and pasting from another program. However, SymPy is not as powerful

[4] SymPy is part of the Anaconda distribution and can be installed with `conda`. See Appendix A for details.

as commercial packages like *Mathematica* or Maple, or the free alternative Sage, which uses Python to bring together over 100 open-source mathematics libraries (`www.sagemath.org`).

8.4.2 Wolfram Alpha

For some purposes, it is easier to turn to a free online resource like `wolframalpha.com`. Go there, and try the examples below. Wolfram Alpha doesn't use the same syntax as Python, but it's not too hard to adapt. All of the examples below can also be executed in *Mathematica* as written.

Integrals

Here is an integral that arises in probability theory:

$$\int_{-\infty}^{\infty} \mathrm{d}x \, \frac{1/\pi}{x^2+1} \frac{1/\pi}{(a-x)^2+1}.$$

Since the value of this integral depends on the value of a, it defines a function of a.

The integral looks daunting, but try entering it into Wolfram Alpha as

```
Integrate[ 1/(Pi^2*((a-x)^2+1)*(x^2+1)), {x, -Infinity, Infinity} ]
```

Notice the syntax used here: The **`Integrate`** function accepts two arguments, enclosed in square brackets. The first is a symbolic representation of the integrand. The second is itself a list (created by the curly brackets) with three entries: The first is the variable to be integrated. The next two specify the limits of integration. Note that predefined functions (**`Integrate`**) and constants (**`Pi`** and **`Infinity`**) are capitalized.

This integral is the *convolution* of two probability density functions. In this case, they are each called Cauchy distributions. Perhaps surprisingly, the result is also of the same general form, though broader than the distributions we started with.[5]

Here is another example. The integral $\int_a^b \mathrm{d}x \, x^n (1-x)^{M-n}$ arises in the study of **credible intervals** for the parameter that defines a Bernoulli trial, given that M trials yielded n successes.[6] We can make progress evaluating it by entering

```
Integrate[ x^n * (1-x)^(M-n), {x} ]
```

This time, the second argument of **`Integrate`** is a list with only one entry, the integration variable; because we don't specify any specific integration range, Wolfram Alpha responds with the indefinite integral, a function of x, M, and n. It may not seem helpful to be told that this function is the "incomplete beta function," but now we can see if Python knows about that function (even if we don't). It does: It is the function **`betainc`**`(a,b,x)` in the **`scipy.special`** module. We can now write code that evaluates this function instead of computing an integral numerically.

If all you need is the normalization integral, that's the case where we integrate from 0 to 1. Wolfram Alpha can evaluate this via

```
Beta[1, 1+n, 1+M-n] - Beta[0, 1+n, 1+M-n]
```

We learn that this integral simplifies to

$$\frac{\Gamma(n+1)\Gamma(M-n+1)}{\Gamma(M+2)} = \frac{n!(M-n)!}{(M+1)!}.$$

---

[5] Nelson, 2015, Chapter 5 discusses Cauchy distributions.

[6] Nelson, 2015, Chapter 6 discusses credible intervals, and this example in particular.

That last expression is easy to include into any code we write that requires the normalization factor of this distribution.

In general, evaluating a special function is faster and more accurate than numerical integration. However, this option is not always available. If you cannot locate a predefined function corresponding to the expression returned by Wolfram Alpha, you may be able to use **quad** to get the numerical result you need. (See Section 6.7.)

Sums

You may have forgotten the result of the infinite discrete sum $\sum_{j=0}^{M} j^3$, but typing

```
Sum[ j^3, {j, 0, M}]
```

into Wolfram Alpha tells you it's $M^2(M+1)^2/4$, a result you may want someday for computing the third moment of a uniform discrete probability distribution.

Ordinary differential equations

The differential equation $\mathrm{d}v/\mathrm{d}t = -Av + B\mathrm{e}^{-ct}$, where A, B, and c are constants, arises in the context of virus dynamics. (See Section 5.1.1.) We can ask Wolfram Alpha to solve it:

```
solve dv/dt = -A*v + B*exp(-c*t), v(0) = vzero
```

To get a definite solution, we need to specify an initial value. In the expression above, we required that $v(t)$ should equal the constant `vzero` at time zero.

Another first-order ordinary differential equation arising in bacterial dynamics[7] can be solved similarly:

```
solve dx/dt = t - x, x(0) = 0
```

[7] See Section 5.2.1 (page 66). Nelson, 2015, Chapter 9 discusses this system and its physical model.

CHAPTER 9

Third Computer Lab

In this lab, you will import images into Python and display them. You will also explore an important operation in image analysis: convolution. If you've ever played with photographic software such as GIMP[1] (or a commercial alternative), you may have used convolutions to smooth or sharpen images but perhaps you didn't know it. Your eyes and brain are also doing convolutions whenever you look at anything.

Not all visual data comes from photographs. Many kinds of experimental measurements generate arrays of numbers associated with points in space. Such data can be communicated to our brains as images, as in atomic force microscopy, computed tomography, and magnetic resonance imaging. Image processing is useful for making these images more meaningful to humans.

Our goals in this lab are to

- Explore the effects of various kinds of local averaging on an image;
- See how to use such averaging to decrease noise in an image; and
- Use specialized filters to emphasize specific features in an image.

9.1 CONVOLUTION

Python gives us access to tools for manipulating images, including convolutions. However, many of the details are hidden. Before getting into Python specifics, let's take a look at the mathematical definition and properties of convolution. A convolution is an operation we perform on probability distributions,[2] but it has other uses as well. A common definition of the two-dimensional discrete convolution C of an image array I with a filter array F is

$$C_{i,j} = \sum_{k,\ell} F_{k,\ell}\, I_{i-k,j-\ell}. \tag{9.1}$$

The sum ranges over all values of k and ℓ that refer to valid entries of both F and I. For example, we cannot have $i = 1$ and $k = 10$ because the first index of I would be -9. (Python may accept this, but the result would not be what you intended.)

Your Turn 9A

a. Consider the trivial transformation, for which F is a 1×1 matrix with a single entry equal to 1. Explain why C is the same as I in this case.
b. Suppose that the size of F is $m \times n$, and that of I is $M \times N$. Explain why the size of C is $(M + m - 1) \times (N + n - 1)$.

---

[1] GIMP is freeware: `www.gimp.org`.

[2] For applications to probability distributions see Nelson, 2015, Chapter 4. For applications to image processing see Nelson, 2017, Chapter 11.

When we convolve an image with a filter, we get another image. The expression in Equation 9.1 is a set of instructions for constructing this new image: To create each pixel in C, we take the pixels from a subset of the original image, multiply them by their respective weights in the filter, and add up the result. It's a simple recipe that can have very different results depending on the filter.

9.1.1 Python tools for image processing

Chapter 8 explained how to load, display, and save images with PyPlot. We also performed a mathematical operation on an array to modify the image. Now we will expand our capabilities by importing a module with several image processing functions:

```
import scipy.ndimage as sim
```

We chose the nickname **sim** as a mnemonic for "SciPy image library."

One useful function in this module is **sim.convolve**. Use **help** to learn about this function and its options. All of the filters we will use in this lab accept similar arguments.

In this lab, you will explore several filters and convolutions. As you proceed through the exercises, you will see a photograph transformed by each operation. To get an idea of what is happening on a pixel-by-pixel level, you can apply the same convolution to a single dot. (This is the **impulse response** of the filter.) This will allow you to see the shape of the filter and better understand what you are doing to the photograph.

Try the following now:

```
impulse = np.zeros( (51, 51) )
impulse[25, 25] = 1.0
my_filter = np.ones( (3, 3) ) / 9
response = sim.convolve(impulse, my_filter)
plt.figure()
plt.imshow(response)
```

We will explain this filtering operation in the next section. For now, compare the size of two arrays, `impulse` and `response`. They are the same! You found in Your Turn 9A that the convolution of an image with a filter is at least as large as the original image, and usually larger. Where do the extra points go in Python? Where did they come from in the mathematical derivation?

Refer back to Equation 9.1 and look at `C[0,0]`. The allowed values of k and ℓ for this point give only one contribution. Likewise, `C[M+m-2,N+n-2]` has only one contribution:

```
           C[0, 0]  is  F[0, 0] * I[0, 0]
C[M+m-2, N+n-2]  is  F[m-1, n-1] * I[M-1, N-1].
```

The points at the edges of the convolved image receive contributions from fewer points in the original image than do the points in the interior. This may lead to distortions at the edges of the convolved image. Python crops the edges and returns only the central portion of the convolved image. This has two advantages. First, every point in the convolved image uses at least a quarter of the filter. Second, convolution does not change the size or shape of the original image.

Even with this cropping, however, points near the edge must be treated differently from those in the center. The function **sim.convolve** and its relatives offer several options. The simplest is to imagine that the image is surrounded by an infinite black border. That means we effectively enlarge our image array to whatever size it needs to be to provide the same number of points for every pixel in the convolution and set the values of all the new points to `0`. To select this behavior, use the keyword and value *mode*=`'constant'` when calling **`sim.convolve`**. This overrides the default behavior, which is to pretend that the image is surrounded by reflections of itself.

9.1.2 Averaging

A very simple filter assigns the same weight to each pixel in a fixed region, as in the impulse response example above. Each pixel in the convolved image is an average of its neighbors in the original image.

Assignment:

Follow the instructions of Section 4.1 to obtain the data set `16catphoto`. Copy the files `bwCat.tif`, `gauss_filter.csv`, and `README.txt` into your working directory.

a. *Make a* 3×3 *array* `my_filter` *in which each element equals* $1/9$. *Why does it make sense to choose the value* $1/9$? We'll call this array the "small square filter."

b. *Use* **`sim.convolve`** *to convolve your new filter with the image you downloaded, and display the result. How does the image change? Try to achieve the same result by using* **`sim.uniform_filter`**.[3]

[*Hints:* You can retain the previous picture for comparison by using **`plt.figure`**`()` to create a new figure before displaying the second. Another option is to display the plots side by side with **`plt.subplot`**. (Consult **`help`**`(`**`plt.subplot`**`)` and Section 4.3.7.) You need to set the color map separately for each figure window that you create. (Consult **`help`**`(`**`plt`**`.colormaps)`.)]

c. *Repeat part (a) using a* 15×15 *array (the "large square filter"). Be sure to use a constant value appropriate to this larger array. How does the image change? How does it compare with the result of the smaller filter?*

d. *Use the definition in Equation 9.1 to show that these convolutions produce images in which each pixel is the average of some of the neighboring pixels in the original.*

9.1.3 Smoothing with a Gaussian

Now we will look at a slightly more complex filter, called a Gaussian filter. Load the file `gauss_filter.csv` into a NumPy array called `gauss`. Then, review Section 6.4.3 to refresh your memory on the function **`plot_surface`**.

Assignment:

a. *Display the convolution of* `gauss` *with the original image. (Also try* **`sim.gaussian_filter`** *with the keyword option* `sigma=5`.*)*

b. *Use* **`plt.imshow`** *to compare the convolutions of a single dot with a Gaussian filter and the square filter you used in question 9.1.2b.*

c. *Use* **`plot_surface`** *to view the convolved images from part (b) in three dimensions.* This is actually a plot of the filters themselves. Use the definition of convolution to explain why. Then, explain how convolution with a Gaussian filter differs from a square filter, and under what circumstances one might prefer a Gaussian filter.

---

[3] The functions **`sim.uniform_filter`** and **`sim.gaussian_filter`** can be significantly faster than **`sim.convolve`** for large filters, but the output is not identical.

9.2 DENOISING AN IMAGE

Measuring instruments, including your eyes, inevitably introduce some randomness, or "noise." You can simulate this effect by making a noisy version of the original image you imported. To do this, multiply each pixel in the original image by a random number.

If you take the default result of **np.random.random**, which is a random variable with mean 0.5, the image will get darker. The image will also no longer be an array of integers between 0 and 255. This can be addressed in three steps:

1. Shift the minimum of the new, noisy image array to zero. That is, find the smallest value of the new array, and subtract that value from every element.
2. Rescale the array so that the maximum value is 255.
3. Change the data type to `uint8`. The **astype** method of an array can perform this conversion:

```
a = a.astype('uint8')
```

Assignment:

a. *Multiply each pixel of the original image by a random number between 0 and 1, then transform the resulting array following the steps above. Compare this noisy image with the original.*

b. *Apply each of the three filters from Sections 9.1.2–9.1.3 (small square, big square, Gaussian) to the noisy image from part (a). Do they improve the image? If so, which one works the best? Why? Zoom in on a small region of the resulting images. How do they compare at this scale?*

9.3 EMPHASIZING FEATURES

You have probably heard news people say "Geeks at NASA's Jet Propulsion Laboratory have enhanced these images. . . ." Let's join the fun.

In between "features" (real things of interest to us) and "noise" (random things), experimental images may contain things that are real, but not of interest to us. We may wish to deemphasize such things, or we may wish to quantify some visual feature. A. Zemel and coauthors encountered such a situation when making fluorescence images of mesenchymal stem cells.[4] When subjected to mechanical stress (stretching), the cells polarize: The internal network of "stress fibers" begins to align in the direction of the stretch. Zemel and coauthors sought to quantify the extent to which the cell was polarized, at every point in the cell.

Follow the instructions of Section 4.1 (page 48) to obtain the data set `17stressFibers`. Copy `README.txt` and `stressFibers.csv` into your working directory.

The data in `stressFibers.csv` differs from that in the image files we used earlier. Look at the maximum and minimum numerical values of the numbers in the array `A`. They lie outside the range 0 to 255. PyPlot will still display the image, and you can perform operations on the array. For consistency, you can rescale the array as described in the preceding section.

The image shows the stress fibers. We will now construct and apply a filter that emphasizes long, slender objects that are oriented vertically.

[4] A fluorescent label for nonmuscle myosin IIa was used to tag the stress fibers. See Zemel et al., 2010.

Assignment:

a. *Execute the following code, then make a surface plot of the filter. Describe its significant features.*

```
# convolution.py
v = np.arange(-25, 26)
X, Y = np.meshgrid(v, v)
gauss_filter = np.exp(-0.5*(X**2/2 + Y**2/45))
```

b. *Use the following "black box" code to modify your filter from part (a), then make a surface plot of the resulting filter. Compare and contrast* `combined_filter` *and* `gauss_filter`.

```
laplace_filter = np.array( [ [0, -1, 0], [-1, 4, -1], [0, -1, 0] ] )
combined_filter  = sim.convolve(gauss_filter, laplace_filter)
```

The matrix `gauss_filter` that you created emphasizes features that are long, slender, and oriented vertically. Other features are averaged out. The array `combined_filter` accentuates the edges of such objects.[5]

c. *Now use* **`sim.convolve`** *to apply the filter to the fiber image, display your results, and comment.*

You may notice that the filtered image has poor contrast. After convolution, the values assigned to some pixels are extremely large or extremely negative, but most points fall in a narrow range between the extremes. Python uses its shades of gray to interpolate between these extremes, giving most points a gray level somewhere in the middle of the range. You can highlight the features you wish to emphasize with another modification:[6]

```
plt.imshow(image, vmin=0, vmax=0.5*image.max())
```

d. *In order to emphasize* ***horizontal*** *objects, repeat the above steps with a different choice for* `gauss_filter`. *Optional: Make two more filters that emphasize objects oriented at* $\pm 45°$ *with respect to vertical.*

---

[5] T2 A Laplace filter emphasizes edges, but is sensitive to noise. A Gaussian filter smooths out noise and edges. Combining the two creates a *Laplace of Gaussian* (LoG) filter that emphasizes edges while suppressing noise. The "elongated" Gaussian used here creates an eLoG filter that emphasizes the edges of objects with a particular alignment.

[6] T2 The keyword arguments implement a "window/level transformation." Pixels whose luminance is less than *vmin* are displayed black, those whose luminance is greater than *vmax* are displayed white, and pixels whose luminance lies between these extremes are displayed in shades of gray. This transformation emphasizes contrast in the range of interest while ignoring features outside that range.

Get Going

Programming ... can be a good job, but you could make about the same money and be happier running a fast food joint. You're much better off using code as your secret weapon in another profession .... People who can code in biology, medicine, government, sociology, physics, history, and mathematics are respected and can do amazing things to advance those disciplines.

—Zed Shaw (2017)

We hope you have enjoyed getting to know Python. You have learned a lot—enough, we hope, for you to continue learning on your own.

You do not have to study Python to learn more about it. Just apply what you have learned as you study subjects that interest you. When you encounter an unfamiliar mathematical concept (perhaps a "wave packet"), create a plot to gain some intuition. When you find an interesting model, explore its behavior as you vary its parameters. When you come across a new statistical concept (perhaps "skewness," "kurtosis," or the "interquartile range"), try applying it to some real or simulated data to learn what it tells you about a collection of numbers. If you do this, then not only will you better understand what you are studying; you will also improve your proficiency with Python. You may need it when it's time for your own research!

For fun and inspiration, we leave you with a script you can run. Give it a minute—literally. This is an example of what you can do on your own now.

```
# surprise.py
import numpy as np, matplotlib.pyplot as plt
max_iterations = 32
x_min, x_max = -2.5, 1.5
y_min, y_max = -1.5, 1.5
ds = 0.002
X = np.arange(x_min, x_max + ds, ds)
Y = np.arange(y_min, y_max + ds, ds)
data = np.zeros( (X.size, Y.size), dtype='uint')
for i in range(X.size):
    for j in range(Y.size):
        x0, y0 = X[i], Y[j]
        x, y = x0, y0
        count = 0
        while count < max_iterations:
            x, y = (x0 + x*x - y*y, y0 + 2*x*y)
            if (x*x + y*y) > 4.0: break
            count += 1
        data[i, j] = max_iterations - count
plt.imshow(data.transpose(), interpolation='nearest', cmap='jet')
plt.axis('off')
```

APPENDIX A

Installing Python

This appendix explains how to install the Python environment described in the main text. The Anaconda distribution is provided and maintained by Anaconda, Inc. It includes the Python language, many libraries, the Spyder integrated development environment (IDE), and the Jupyter Notebooks system in a single package. (A software **package** is a collection of programs and data files related to a particular application or library. The data files include information about any other packages that may be required for the new software to function properly.) Anaconda also provides a simple protocol for updating packages and installing additional packages. There are many alternatives available. You may find one better suited to your own needs or preferences. One place to look is `www.python.org`.

Our intention is not to promote a particular distribution of Python, but to promote Python itself as a tool for scientific computing.

A.1 INSTALL PYTHON AND SPYDER

You can download Anaconda at `anaconda.com/download`.

You must first choose your installer. The site automatically detects your operating system, but do not download anything just yet. First, make sure the installer offered to you matches your operating system: macOS, Windows, or Linux. Second, make sure your operating system meets the requirements for the current version of Anaconda. You can find the current requirements at `docs.anaconda.com/anaconda/install`. Third, make sure you choose the proper version of Python. The instructions below describe how to install the current version of Python 3 at the time of this writing. (Newer versions of Anaconda and Python will presumably follow a similar installation process, but consult the online documentation at `docs.anaconda.com` if you have trouble.) Finally, you must decide which type of installation program you want: graphical or command line. The graphical installation (point and click) is easiest, unless you already enjoy working with your operating system's command line.

If your computer meets the system requirements and you wish to carry out a graphical installation, proceed to Section A.1.1.

If your operating system does *not* meet these requirements, the standard installation may not succeed on your computer, or the installation may succeed, but some packages may not work properly. You might wish to try Section A.1.1 anyway. If the installation doesn't work, it is easy to undo the process by uninstalling Anaconda. Another option is to try installing from the command line as described in Section A.1.2. If that also fails, you can download an older version of the Anaconda or Miniconda installer:

Anaconda `repo.continuum.io/archive`
Miniconda `repo.continuum.io/miniconda`

Try matching the approximate date of your operating system with the release date of the installer for best results.

The Anaconda website offers extensive documentation, so even if the instructions given here are insufficient, you should still be able to get Anaconda up and running on your computer. For the most detailed and up to date installation instructions, go to `docs.anaconda.io/anaconda/install`.

A.1.1 Graphical installation

macOS

To get the latest stable version of Python (currently Python 3.6), on `anaconda.com/download/` find the "Download" button directly beneath "Python 3.6 version." Click on the link to download the package. This is a large download that will install many packages not used in this tutorial; however, the installation is very easy. Unless you are comfortable working from the command line, this is the installation we recommend.

Once the download finishes, find the package and open it. It is probably in your `Downloads` folder. Double-click on the file to initiate an installation dialog. Unless you want to customize the installation, just click "Continue" until you reach "Destination Select." Select "Install for Me Only" unless you have a good reason for doing otherwise. This will create a folder called `anaconda` in your home directory and store all the associated files there. Click "Continue" to bring up the "Installation Type" dialog. Click "Install" to initiate a standard installation. You may be prompted for a password.

Once the installation is complete, you may find a new icon on your Desktop called `Anaconda Navigator.app`. Open this application. Click on the LAUNCH button below "spyder" to start Spyder. You can also start Spyder without `Anaconda Navigator.app` by double-clicking on `spyder` in the folder where Anaconda installed it. By default, this is `anaconda/bin`. A third option is to start Spyder from your operating system's command line by typing

```
$ spyder
```

and hitting `<Return/Enter>`. (Don't type the $.)

Once you have successfully installed Anaconda and launched Spyder, proceed to Section A.2.

Windows

To get the latest stable version of Python (currently Python 3.6), on `anaconda.com/download/` find the "Download" button directly beneath "Python 3.6 version." Click on the link to download the package. This is a large download that will install many packages not used in this tutorial; however, the installation is very easy. Unless you are comfortable working from the command line, this is the installation we recommend.

Once the download finishes, find the package and open it. It is probably in your `Downloads` folder. Double-click on the file to initiate an installation dialog. Unless you want to customize the installation, just click "Continue" until you reach "Destination Select." Select "Install for Me Only" unless you have a good reason for doing otherwise. This will create a folder called `anaconda` in your home directory and store all the associated files there. Click "Continue" to bring up the "Installation Type" dialog. Click "Install" to initiate a standard installation. You may be prompted for a password.

Once the installation is complete, you may find a new icon on your Desktop called `Anaconda Navigator.app`. Open this application. Click on the LAUNCH button below "spyder" to start Spyder. You can also start Spyder without the Desktop application. After the install, you will find a new folder called `Anaconda` or `Anaconda (64-bit)` under `Start>All Programs`. This folder contains a shortcut to launch Spyder. The script that launches Spyder may also be found in the User directory under `Anaconda\Scripts\spyder-script.py`; alternatively, `Anaconda\Scripts\spyder.exe` will also work.

Once you have successfully installed Anaconda and launched Spyder, proceed to Section A.2.

A.1.2 Command line installation

To install Python on a Linux machine, or to install a leaner environment that includes only the packages you want under macOS or Windows, you can use the `conda` package manager. You can directly access this tool from the command line.

In this appendix, "command line" refers to a command line interface for your operating system—not the IPython command prompt. For macOS users, this is `Terminal.app`. For Windows users, it is `cmd.exe`. For Linux users, it is a bash shell. We have indicated the command prompt as `$`, but it may look different on your system. Do not type the command prompt symbol when entering the following commands.

To install Anaconda from the command line, first, download Miniconda. Go to `conda.io/miniconda` and select the current version of Python 3 for your operating system.

Once the download is complete, macOS and Linux users will need to run a script. At the command line, type

```
$ bash ~/Downloads/Miniconda3-latest-MacOSX-x86_64.sh
```

for macOS, or the equivalent command for your operating system, file name, and file location. This will run the script. Windows users should double click the `.exe` file and follow the instructions on the screen. After you agree to the license terms and select a directory for the installation (or agree to the default location), the script will install Python 3 and the `conda` package manager. (Note that you need to specify the complete path: `/Users/username/anaconda` or `/anaconda` for macOS or Linux, or `C:Anaconda` for Windows. If you only provide the name "`anaconda`," then files will be installed in a new directory named `anaconda` in your current directory, wherever that may be.)

If this was successful, you can now run Python, but you will not yet be able to use modules like NumPy, PyPlot, or SciPy, and you will not have access to the Spyder IDE. You can now install them, and the entire Anaconda Python distribution, by typing

```
$ conda update conda
$ conda install anaconda
```

If you prefer to install individual packages to minimize disk space, this is simple using `conda`: Instead of the second line above, use commands like

```
$ conda install numpy matplotlib scipy ipython
```

If you watch the screen after you execute these commands, you will see what `conda` does. It determines which packages are necessary to install the specific ones you asked for, downloads all of the required packages, and then installs and links everything.

Python comes with the IDLE integrated development environment. However, if you want to use the Spyder IDE described in this tutorial, you will need to install it.

```
$ conda install spyder
```

In principle, this should install all the packages you need to run Spyder. Launch Spyder by typing

```
$ spyder
```

In practice, you may find that some modules are missing and Spyder will not run. Spyder is actually a colossal Python program. When something is amiss, it displays Python error messages. Perhaps Spyder crashes when you try to run it. At the end of the terminal output you see

```
ImportError: No module named 'docutils'
```

Just issue the command

```
$ conda install docutils
```

and this problem is fixed. To get Spyder up and running, install any other missing modules using the `conda install` command.

Even after Spyder is running, you may not have access to all of the debugging tools described in this tutorial. If you are writing your own code later and find another missing module, you may be able to install it with `conda`.

To run all of the code samples in this tutorial, you will need to use `conda` to install the following packages:

`conda install ipython`	IPython interpreter
`conda install numpy`	NumPy
`conda install matplotlib`	Matplotlib and PyPlot
`conda install scipy`	SciPy
`conda install pillow`	Image processing library
`conda install sympy`	Symbolic computation library

To use the Spyder IDE or Jupyter notebooks, you will need to install them separately:

`conda install spyder`	Spyder IDE (discussed above)
`conda install jupyter`	Jupyter Notebooks

After installing Anaconda, you can run `anaconda-navigator` and `spyder` from the command line. If you use Miniconda instead, you can install `anaconda-navigator` and `spyder` separately, and then run these programs from the command line.

This set of packages should be sufficient to complete the exercises in this tutorial. However, there is a lot more you can do with `conda`. For example, you can set up environments that allow you to switch between Python 2 and Python 3, or use different versions of NumPy and SciPy. You can learn more about `conda` at `conda.io/docs/`.

A.2 SETTING UP SPYDER

Now that you have installed Spyder, there are a few settings to fine tune before you start writing and executing scripts. All of these finishing touches will be made using the Preferences. To access the Preferences, click the wrench icon, or select it from a drop-down menu: `Tools>Preferences` on Windows, or `spyder>Preferences` on macOS. After making the changes below, you may need to restart Spyder before they take effect.

A.2.1 Working directory

You will need to keep track of your files. The easiest way to do this from within Spyder is to tell it to save all of your work in a folder that you specify. Choose "Global working directory" from the list of options on the left of the Preferences panel. On the right, choose the button next to "the following directory," and then click on the folder icon to select a directory, or enter the path of the directory you wish to use. You may wish to create a new folder named `scratch` or `current` to work in. After you have selected a folder, click the appropriate buttons so that "Open file" and "New file" use this directory.

You can still access files anywhere on your computer, but setting these options is the easiest way to locate the files you create with Spyder.

A.2.2 Interactive graphics

Spyder's default option is to display graphics in the IPython console, similar to programs like Maple and *Mathematica*. However, these plots are static images. You cannot zoom, move, or rotate these plots. To make interactive plots the default, click the menu item `python>Preferences` and select the "IPython console" tab on the left of the Preferences window. Click on the GRAPHICS button at the top of the menu on the right. In the panel called `Graphics backend`, set `Backend` to `Automatic`, `Qt5`, or something other than `Inline`. Finish by clicking APPLY at lower right.

You may have to restart Spyder before this change takes effect.

A.2.3 Script template

It is good practice to include the author, creation date, and a brief description in any scripts you write. You will also import the NumPy and PyPlot modules in nearly every script you write for this tutorial. You can give all of your scripts a standard header by using the Preferences.

Choose the menu item `python>Preferences` and select the "Editor" tab on the left of the Preferences window. Next, click on the ADVANCED SETTINGS button at the top of the menu on the right. Click on the button marked EDIT TEMPLATE FOR NEW MODULES. This will open a file in the Editor. Make changes to this file so that it contains text you want to include in every script you write. Here is a sample:

```
# -*- coding: utf-8 -*-
# file_name.py
# Python 3.6
"""
Author:      Your Name
Created:     %(date)s
Modified:    %(date)s

Description
-----------
"""
import numpy as np
import matplotlib.pyplot as plt
```

The expressions with percent signs will insert the date when a file is created. The first line in this sample template is optional. You should include it if your code makes use of any characters not in the "standard ASCII set," such as non-English characters and accents.

After you have finished editing this file, save your changes and close it.

A.2.4 Restart

After you make these changes and adjust the other preferences to your liking, quit Spyder and relaunch. You should now be ready to carry out any of the exercises in this tutorial.

A.3 KEEPING UP TO DATE

In the future, there will be new releases of Python and its modules, and the people in charge of the Anaconda distribution will work to eliminate bugs. You can get the latest and greatest Anaconda distribution by using the `conda` package manager. This can be done from your operating system's command line.

At the command line, type

```
$ conda update conda
$ conda update anaconda
```

The package manager will determine which files need to be downloaded, installed, and updated. You can choose to accept or reject the proposed changes, and `conda` will do the rest.

You can update or install individual packages with `conda`. For example, `conda update numpy` will download and install the latest version of NumPy.

The Anaconda Navigator also allows you to update or install individual packages. It provides a graphical interface to the `conda` package manager. Click on the `Environments` tab on the left side of the navigator window. There are drop-down menus and buttons that allow you to manage the packages in your installation. To update a package you have already installed, select `Installed` from the first drop-down menu; to install a new package, select `Not installed` from the first drop-down menu. Locate the package you wish to update or install and click the `Update index ...` button. Then `conda` will take care of the rest.

For more information on `conda`, see `conda.io/docs`.

A.4 INSTALLING FFMPEG

FFmpeg is an open-source encoder for creating audio and video files. It can be used to create animations within Python, as described in Section 8.3. However, FFmpeg is not part of Python, nor is it available as part of the Anaconda distribution. The easiest way to install it on your computer is to use a package manager. This is a program that will determine all the libraries that are necessary to run the program you request, then download, install, and link everything together so your program runs properly. `conda` is a package manager. MacPorts (`macports.org`) and Homebrew (`brew.sh`) are popular options in macOS. In Linux, `yum` (`yum.baseurl.org`) and `apt-get` (`wiki.debian.org/Apt`) are widely available. The procedure for installing FFmpeg is similar for all of these, but for simplicity, we will describe the procedure for `conda`.

Many individuals and organizations that are not associated with Anaconda, Inc. have made large repositories of code available through the `conda` package manager. A repository like this is called a "channel" and we need to direct `conda` to the appropriate channel to install FFmpeg. (This example makes use of the conda-forge channel: `conda-forge.org`.) Using a channel requires a slight modification to the installation command:

```
conda install --channel conda-forge ffmpeg
```

The installation should proceed as with other Python packages, even though FFmpeg is not a Python package. Once `conda` has finished, you can test the installation by asking your operating system to locate the program `ffmpeg`. Type

```
$ which ffmpeg
```

at the command line. You should see something like "`/Users/username/conda/bin/ffmpeg`." This indicates that Python will now be able to find and run FFmpeg.

A.5 INSTALLING IMAGEMAGICK

ImageMagick is a collection of open-source command line tools for creating and manipulating images. It can also be used to create animations within Python, as described in Section 8.3. Unlike FFmpeg, ImageMagick is not universally available through the `conda` package manager.

Linux users can install ImageMagick using `conda`, but from a different channel:

```
$ conda install --channel kalefranz imagemagick
```

You can also download and install ImageMagick by following the instructions at `imagemagick.org`.

However, just to show that using a package manager need not be intimidating, we close by illustrating the use of the Homebrew package manager for macOS to install FFmpeg and ImageMagick.

1. Install Homebrew by following the instructions here: `brew.sh` (It only requires copying and pasting a single line of code into the command line of `Terminal.app`.)
2. Open `Terminal.app`. At the command line, type

   ```
   $ brew install ffmpeg imagemagick
   ```

3. Wait a couple of minutes while Homebrew installs the programs.
4. Verify the installation. Type the following at the command line:

   ```
   $ which ffmpeg
   $ which magick
   ```

If your operating system was able to locate these programs, you should be able to run all of the animation codes in Section 8.3.

APPENDIX B

Jupyter Notebooks

There are many ways to write and run Python code. You could simply to write scripts with a text editor and run them from your operating system's command line. Many "front end" systems have been developed to make coding easier. One of these is the Spyder integrated development environment (IDE) described throughout the main text. This appendix describes another popular choice: the Jupyter Notebooks system.[1]

Like a physical notebook, a Jupyter notebook is a record of your work. It is an interactive document that is displayed in a web browser. However, all files are stored on your own computer, not on a web server. Each notebook consists of a series of **cells**. Cells may contain either executable code or formatted text. This means you can include presentation-quality documentation alongside your source code. Jupyter notebooks are also easy to share. Another person can open your notebook and run or modify the code while consulting your commentary.

Let's look at a Jupyter notebook now.

B.1 GETTING STARTED

Although a Jupyter notebooks opens in your web browser, the software must be installed on your computer. Jupyter Notebooks is included with the Anaconda distribution. It can also be separately installed with the `conda` package manager. (See Appendix A.)

B.1.1 Launch Jupyter Notebooks

Once you have installed the software, you can launch a Jupyter Notebooks session from your operating system's command line with the command

```
$ jupyter notebook
```

You can also use the Anaconda Navigator application to launch the Jupyter Notebooks application. (You may get a message like

```
Copy/paste this URL into your browser when you connect for the first time,
to login with a token: http://localhost:8888/?token=XXXX
```

Do as you are instructed to get started.)

A window will open in your default web browser with "Jupyter" at the top left corner (Figure B.1).[2] The first of three panes is selected by default. This is a file browser that you can use to find and

[1] "Jupyter Notebooks" is an extension of "IPython Notebooks" that will eventually be subsumed into "JupyterLab." The interface is similar to the notebook environment of *Mathematica*, Maple, and Sage.

[2] T2 Starting Jupyter launches several "background processes" that you do not see. These processes communicate with your web browser, which communicates with you. Both Jupyter and your notebooks are "local" since they are located on your computer. The web browser simply provides a convenient and familiar interface for human users.

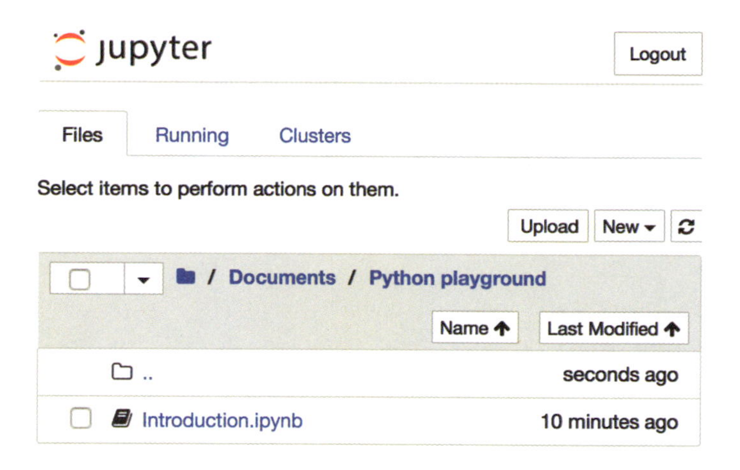

Figure B.1: The Jupyter Notebook file browser.

open an existing notebook or create new notebooks. If you open a Jupyter notebook from your operating system's command line, the file browser will start in the directory from which you issued the command. If you launch from Anaconda Navigator, the file browser will start in your home directory.

B.1.2 Open a Notebook

Since you don't have any notebooks yet, use the file browser and navigate to a folder where you'd like to create one. (You can create a new folder with the [New] button at the top right of the file browser. Check the box next to the new folder and then click [Rename] to rename it.) Next, click on the [New] button and select "Python 3" from the dropdown menu. A new tab will open with a Jupyter notebook that contains a single empty code cell called `In [ ]` (Figure B.2). Enter `1+1` and then click the ⏯ button or hit `<Shift-Enter>` on your keyboard. You have just run your first code in a Jupyter notebook!

You can save your notebook with the 💾 button, the menu item `File>Save and checkpoint`, or the shortcut `<Cmd-S>`. You can now find it in your computer's file system (probably named `Untitled.ipynb`). If you would like to rename a file, you can simply click on the name of the notebook at the top of the page. You can also use the menu option `File>Rename...`.

You can also open existing notebooks saved on your computer. For example, you can view a sample notebook at the blog that accompanies this book.[3] It illustrates many features of Jupyter notebooks. You can download the notebook, open it, modify it, and play with your own copy. Many other notebooks are also available for download from the web.

B.1.3 Multiple Notebooks

You can run multiple notebooks at the same time. Return to the file browser tab, create a second new notebook, and perform some other calculations. You now have two independent notebook sessions. Each one has its own state, including the values of all variables. No information is shared between

[3] `physicalmodelingwithpython.blogspot.com/2016/01/jupyter-notebooks.html`.

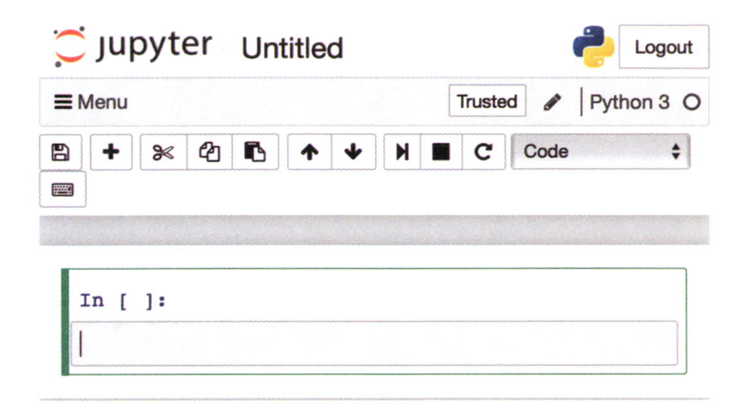

Figure B.2: A new Jupyter notebook. The green border on the selected cell indicates that Jupyter is in input mode.

notebooks. You can keep track of all open notebooks in the file browser tab: just select the RUNNING pane. From this pane, you can selectively terminate any of your running notebooks.

B.1.4 Quitting Jupyter

When you are finished computing and editing, you will need to be sure you terminate all of the processes Jupyter is running on your computer. Simply closing the tabs in your web browser will not do the trick.

First, shut down each notebook. You can do this from within a notebook with the menu option `File>Close and Halt`. This will close the notebook and shut down the Python process that was running in the background. Alternatively, you can use the file browser. Select the "Running" pane to see a list of all active notebooks. There will be a SHUTDOWN button next to each. Use it to shut down all active notebooks.

Next, you need to terminate the Jupyter Notebooks application itself. To do this, first locate the "terminal window" associated with the application. If you launched the Jupyter Notebooks application from the Anaconda Navigator, you will need to find an application window that displays a series of messages from `NotebookApp`:

```
[I 03:14:04.679 NotebookApp] ...
[W 03:17:26.088 NotebookApp] ...
...
```

If you started Jupyter Notebooks from your operating system's command line, this will be the terminal window where you entered the startup command. Click on this application. Close the Jupyter Notebooks application once and for all by typing `<Ctrl-C>`. The program will ask for confirmation. Press `<y>` then `<Enter>` to confirm. The program is now closed.

The remainder of this appendix provides a brief overview of some useful features of Jupyter notebooks. However, the best way to learn how to use Jupyter notebooks is to create your own and explore those created by others.

B.1.5 T2 Setting the Default Directory

It is good practice to keep all of your work in one location. You can set the default starting folder for Jupyter notebook sessions. (This is analogous to choosing the "Global Working Directory" in Spyder.) However, this is not currently possible from within Jupyter. You must use your operating system's command line. At the command line, type

```
$ jupyter notebook --generate-config
```

This will generate a configuration file called `jupyter_notebook_config.py`. This is simply a Python file that sets a lot of options every time Jupyter starts up. It will be located in a hidden folder called `.jupyter` in your home directory. Open this file in a text editor. Search for the lines

```
## The directory to use for notebooks and kernels.
#c.NotebookApp.notebook_dir = ''
```

Uncomment the second line and replace the empty string with the name of the folder where you want Jupyter to run, such as `"/Users/username/scratch/jupyter"`. Save and close. The next time you start Jupyter, it will open in this folder.

B.2 CELLS

When you created a new, empty notebook, it consisted of a single cell. Every Jupyter notebook is composed of one or more cells. Each cell may be one of the following three types:

- Executable Python code (or code in more than 40 other programming languages), possibly accompanied by output from that code
- Text formatted with the Markdown system
- Plain text

You can select any cell and then insert a new one before or after it by using the `Insert` menu.

B.2.1 Code cells

Code cells are those in which you enter commands for Python. A single code cell may contain multiple Python statements. If the code generates text output, it will appear below the input after an `Out [N]:` prompt.

A series of code cells resembles a Python script, but Python will not necessarily carry out commands in the order you enter them. When you create a new notebook or open an existing one, Jupyter launches an IPython **kernel** in the background. This is the program that executes all of the Python code during your session.[4] You enter your code in code cells, and you can run these in any order you choose. You can also select a group of cells and run them all in sequence by using the `Cell` menu.

The notebook informs you of the order in which cells were evaluated. When you created an empty notebook the cell was labeled `In [ ]`. When you asked Jupyter to evaluate `1+1`, the name of the code cell changed to `In [1]` to indicate that this was the first cell you evaluated in that session. When

[4] Jupyter Notebooks and the Spyder IDE are built on top of the same IPython interpreter. All the Python code and IPython magic commands in this book can be run in code cells.

you have many cells, they can be run in any order you wish. The cell name indicates the sequence number of the most recent evaluation of that cell.[5]

Each notebook remembers its state as you execute various cells. Even if you delete a cell from the notebook, that cell's effect on the kernel's state persists. To start fresh, you can restart the kernel. Use one of the restart options available from the `Kernel` menu at the top of the page. There is also a ⟳ button for this purpose.

After you terminate or restart a kernel, all of your output remains visible in the code cells, even though the state of the kernel has been lost. That's good if you want to save the notebook or share it with another reader. However, this behavior can be confusing if you want to rerun your notebook from scratch. You can easily get confused about which output came from the latest run and which came from the previous one. For this situation, there is a convenient menu item called `Kernel>Restart & Clear Output`.

B.2.2 Graphics

If the code in a cell generates a figure, that will appear in the output region of the cell by default. (You may need to add a `plt.show()` command to render graphics.)

If you prefer an interactive graphic instead of the default "inline" behavior, include the following "magic" command immediately after importing PyPlot:

```
%matplotlib notebook      # Create interactive graphics.
```

(This should eliminate the need for any `plt.show()` commands.) New figures have a "power button" on the upper-right corner of the interactive plot window. Do all of your adjustments and explorations before you click on it. Once you click on it, the figure will be frozen in its current state until you run the cell to generate it again.

If you want to switch back to inline graphics, use the magic command

```
%matplotlib inline        # Display static output in the cell.
```

B.2.3 Markdown cells

New cells are code cells by default, but you may wish to add some explanation, commentary, or reflection. You could use Python comments within code cells, but Jupyter notebooks offer a more flexible method for entering text. Select one or more cells and use the `Cell>Cell Type` menu to change them to Markdown mode. Now, you can type text in the cell and it will not be executed. Your typing shows up as plain text as you type. When you are finished, evaluate the cell with the ⏭ button, as you would a code cell. The plain text input is replaced by more aesthetically pleasing output. If you want to change what you have typed, double-click on the cell. You will see the plain text that you entered, and you can begin editing.

You can format text using the Markdown language. Cells can even contain mathematical expressions rendered with MathJax.[6] A description of these features is beyond the scope of this appendix, but the sample notebook demonstrates some of the possibilities. A web search for "`jupyter notebook gallery`" provides many more examples.

---

[5] When a cell takes a long time to run, its name temporarily changes to `In[*]` to indicate that it is still being processed.

[6] MathJax allows you to use most TeX and some LaTeX commands to produce beautiful mathematical output in HTML. See `docs.mathjax.org`.

B.2.4 Edit mode and command mode

A Jupyter notebook has two primary modes of operation. In command mode, the keys you type carry out operations within the notebook: saving text, evaluating code cells, inserting or deleting cells, and so on. In edit mode, the keys you type produce text within a cell.

You can enter edit mode by double-clicking on any cell. If a cell is highlighted, you can also enter edit mode simply by pressing <Return/Enter>. (Keep in mind, however, that editing a cell has no effect on the kernel's state until you actually run the cell.) You can leave edit mode and enter command mode at any time by clicking outside of a cell, or by pressing the <Esc> key. Now many common operations can be carried out with a single keystroke. For example, <X> will delete the current cell. If you accidently press <X> while in command mode, you can undo the operation with <Z>. Another useful shortcut is <S>, which saves your notebook. See Help>Keyboard Shortcuts for the complete list. Learning a few keyboard shortcuts can make using Jupyter notebooks a lot easier and more efficient.

B.3 SHARING

You can share a notebook file with anyone who also runs Jupyter just by sending them the .ipynb file. You can also export a static version in several other formats. An HTML version of your notebook requires nothing but a web browser to view it. If you have the free LaTeX software installed on your system, you can also convert the notebook to a PDF file. Finally, you can share your notebooks over the web.

If you save a notebook to Dropbox, you can share a link to the notebook with others. They can go to the Jupyter notebook viewer (nbviewer.jupyter.org), enter the link, and view your notebook—even if they do not have Python or Jupyter Notebooks installed on their computer. The code repository github.com even has a notebook viewer built into the website. If you upload your notebook there, anyone can view it, download it, or use a link to the file to view it in a notebook viewer.

B.4 MORE DETAILS

The information in this appendix should be enough to get you started with Jupyter Notebooks. The notebook Introduction.ipynb available on this blog that accompanies this book demonstrates more techniques. For more information, have a look at the official documentation: jupyter-notebook.readthedocs.io. That site contains information on Markdown, MathJax, other programming languages, and many other topics. Much of this information is even available in Jupyter notebooks.

B.5 PROS AND CONS

Jupyter notebooks allow you to *do* your work, document your work, and see the results in a single file. This makes them very useful for exploring new ideas, solving problems, and collaborating. Jupyter notebooks are also useful in presenting your work. The notebook also eliminates the often painful step of including code and graphics in a formatted document. (Try creating a similar document with a word processor, and you will see the advantages of Jupyter notebooks!) Jupyter notebooks are also useful in teaching. A teacher can use notebooks for demonstrations, in-class collaborative work, and assignments. A partially completed notebook can be a useful exercise for students.

Jupyter notebooks do many things very well. However, no tool is perfect for every job. The browser interface places another layer of software between the programmer and the machine, and performance can suffer. Complex graphics can overwhelm a browser at times. Many scientists and programmers prefer an IDE such as Spyder for writing and debugging large programs—especially those involving multiple files. It can be difficult to transform a notebook into a polished piece of software.

Give Jupyter notebooks a try. You may find them to be the perfect tool for some of your work.

APPENDIX C

Errors and Error Messages

Now would be a good time to start making errors. Whenever you learn a new feature, you should try to make as many errors as possible, as soon as possible. When you make deliberate errors, you get to see what the error messages look like. Later, when you make accidental errors, you will know what the messages mean.

— Allen B. Downey

Everyone makes mistakes. When Python detects something wrong, it will halt a program and display an error message. This process is called "raising an exception."

Of course, Python doesn't really know what you are trying to accomplish. (At least not yet ...) If you enter something syntactically correct, but not what you need to solve your problem, Python can't point that out to you. Hence the importance of testing your code and debugging it. Even when Python does detect something wrong, it may not be able to tell you what you need to fix.

This appendix makes no attempt to catalog all of Python's error messages. Instead, we show a few common errors and decode them. As you learn about how Python interprets your code, you'll gain insight into these messages.

C.1 PYTHON ERRORS IN GENERAL

Before getting down to specifics, let's take a look at Python's general method for handling errors. Follow the advice at the beginning of this chapter and make a few errors. First, type `%reset` to erase all variables and modules. Then, type the following commands at the command prompt as written. (Feel free to make your own errors as well.)

```
1/0
import nump as np
abs('-3')
a = [1,2,3]; a[3]
b = [1,2,3)
b[0]
```

You should get a different error message from each command. Notice that each error has a name—`ZeroDivisionError`, `TypeError`, `SyntaxError`, and so on—followed by a message. When Python detects an error, it raises an exception. You can raise an exception manually, and even provide your own message. Try these commands:

```
raise TypeError
raise ZeroDivisionError("You should not divide by zero!")
```

Python has classified many common errors. Thus, when it recognizes an error, it is able to describe what went wrong and provide additional messages supplied by programmers. These messages may not seem very informative at first, but compared with the "segmentation fault" of lower-level programming languages, they are quite helpful.

Python not only allows programs to raise exceptions; it also allows you to catch them before they terminate your program. To see the difference in behavior, try the following two methods of dividing by zero:

```
import numpy as np
np.divide(1, 0)
1/0
```

The programmers who created NumPy decided that sometimes it is OK to divide by zero because some mathematical functions do have singularities. So they downgraded the exception from **ZeroDivisionError** to **RuntimeWarning**, thus allowing some singular functions to return a value.[1] This behavior allows NumPy to process an entire array and return a result, even if some of the entries involve operations that are not well defined mathematically. In contrast, basic Python stops execution and alerts the user.

You can override the default behavior by using Python's protocol for handling exceptions. A function call may raise an exception, but you can catch it before it halts the entire program and do something else. Try typing

```
try:
    1/0
except ZeroDivisionError:
    print("1/0 -> infinity")
```

You do not need to master exceptions and exception handling at this point. The takeaway message from this section is that Python will try to alert you to the nature of the error by raising an exception, but not every program will handle the same errors in the same way.

Now let's take a look at some common errors.

C.2 SOME COMMON ERRORS

- **SyntaxError** – This is the most common error for beginning programmers. It usually means you typed a command incorrectly. To generate a syntax error, try the following:

```
abs -3
```

It may be obvious to a human reader what you mean here, but Python is not a human reader. It knows of a function called **abs**, but this function should be called with a single argument in parentheses. You did not call this function using the proper syntax, so Python raises a **SyntaxError**.

Sometimes Python is able to pinpoint exactly where the error occurred. Suppose that you used a single equals sign when you wanted to test for equality:

```
if q = 3: print('yes')
else: print('no')
```

[1] Some IDEs will suppress the **RuntimeWarning** and simply return the value **np.inf** without complaining.

Python replies with

```
File "<ipython-input-87-19c154aec9ce>", line 1
  if q = 3:
       ^
SyntaxError: invalid syntax
```

Here the caret ("^") pinpoints where Python's scanner had arrived when it detected something wrong. A syntax error may actually be located in a line *preceding* the one that is flagged. Python waits until it is "sure" there is an error before issuing the exception. If you find that a flagged line looks correct, work backward to see if there is an error on an earlier line.

Unmatched parentheses are a common source of syntax errors. For example, try the following:

```
x = -3
print(abs(x)))
```

Python will also raise a **SyntaxError** if it reaches the end of a file while still looking for arguments to a function, closing parentheses, or something similar.

```
x = -3
print(abs(x)
```

If you type this in a script and then attempt to run the script, you will get a **SyntaxError**. However, if you type these commands at the command prompt, IPython won't allow you to commit the error. Every time you hit `<Return/Enter>`, the cursor moves down a line and nothing else happens. Just close the parentheses, hit `<Return/Enter>`, and Python will execute the command.

- **ImportError** – Python raises this exception when it cannot find a module you are attempting to import or when it cannot find the function or submodule you are attempting to import from a valid module. Most often, this occurs when you type the name of the module or function incorrectly. (Remember, Python names are case sensitive.) Each of these lines raises an exception:

```
import NumPy
import nump as np
from numpy import stddev
```

If you are certain a module exists and you are spelling its name correctly, then you may need to install the module on your computer, perhaps by using the `conda` tool. (See Appendix A.) If you have installed the module, you may need to move it to a different directory or modify your computer's `PYTHONPATH` environment variable in order for Python to find it.

- **AttributeError** – Python raises this exception when you ask for an attribute or method from an object that does not possess it:

```
np.atan(3)
np.cosine(3)
```

This is usually due to spelling the name of an attribute or method incorrectly, or misremembering its abbreviation.

- **NameError** – Python raises this exception when you ask for a variable, function, or module that does not exist. This can result from misspelling the name of an existing variable, function, or module. It can also happen when you forget to import a module before using it.

```
%reset
zlist
np.cos(3)
```

• **IndexError** – Python raises this exception when you provide an index that lies outside the range of a list or array:

```
x = [1, 2, 3]
x[3]
```

You may make this error a few times before you get used to Python's system of numbering elements starting at 0 rather than 1.

• **TypeError** – Python raises this exception when you call a function with the wrong type of argument. This can occur if you asked for user input and forgot to convert the string to a number before performing a mathematical operation. It can also occur if you try to combine two objects in a way Python cannot interpret:

```
abs('-3')
x = [1, 2, 3]; x[1.5]
2 + x
```

• **AssertionError** – Python raises this exception when an assertion statement is violated. It may be the most helpful error message of all, because *you* ask Python to alert you to the problem.

```
x = [1, 2, 3, 4]
y = [1, 4, 9, 16, 25]
assert len(x) == len(y), "Lists must be same length!"
```

Python has more than 40 built-in exceptions, each for a unique type of error. See how many you can generate!

APPENDIX D

Python 2 versus Python 3

There are two major versions of Python in wide use. In 2008, version 3.0 of Python was released. This version was the first to break *backwards compatibility*, meaning that code written in earlier versions of Python was not guaranteed to run in version 3.0 and later.

In this tutorial, we have chosen Python 3 when a choice was required. If you are already familiar with Python 2 and do not want to change, have no fear. All the modules we describe are available in both Python 2 and Python 3, and we have taken care to write code that will work in either version whenever possible. There are only three instances where accommodating both versions was not practical: division, the **print** command, and user input.

Fortunately, there is a special module called `__future__` that makes many features of Python 3 available within Python 2. This includes both division and the `print()` function. To run all of the code samples in this tutorial in Python 2.7, you only need to add the following two lines at the top of each script and execute them at the beginning of each interactive session:

```
from __future__ import division, print_function
input = raw_input
```

To understand why, read on. To try out code samples, you can return to the main text now.

D.1 DIVISION

In Python 2, division of two integers returns the quotient and ignores the remainder. Thus, the result is always an integer, even if the integers are not evenly divisible. In Python 3, division of two integers instead returns a floating-point number if they are not divisible.

```
1/2 == 0                        # True in Python 2; False in Python 3
1/2 == 0.5                      # False in Python 2; True in Python 3
```

The second option is usually what we want in numerical computation. Users of Python 2 can get this behavior from the `__future__` module:

```
from __future__ import division
```

Integer division is still available in both versions of Python:

```
3//2 == 1                       # True in Python 2; True in Python 3
```

D.2 PRINT COMMAND

The **print** command behaves differently in Python 2 and Python 3, though this will not have a significant effect on the code samples in this tutorial.

In Python 2, **print** is a statement, like **assert**, **for**, **while**, and others. The command

```
print "Hello, world!"
```

will do what you expect in Python 2, but in Python 3 it will raise a **SyntaxError**. That's because **print** is a *function* in Python 3. It requires arguments enclosed in parentheses, like every other function.

In this tutorial, all print commands have the form required by Python 3:

```
print("Hello, world!")
```

This particular statement will also work in Python 2. However, if you use Python 2, be aware that it will interpret a series of arguments in parentheses as a tuple. Thus,

```
print('x', 'y', 'z')
```

will produce different output in Python 2 and 3.

You can bring the Python 3 **print** function into Python 2 by importing it from the `__future__` module:

```
from __future__ import print_function
```

There are two caveats. First, you cannot have both variants of the **print** command at the same time.

```
print "Hello, world!"
```

will now result in a **SyntaxError**. Second, importing from the `__future__` module is irreversible. You must restart Python to revert to standard Python 2 behavior.

D.3 USER INPUT

Python 2 has two statements for obtaining input from the user: **input**`()` and `raw_input()`. In contrast, Python 3 has only one: **input**`()`. However, Python 3's **input**`()` function behaves like Python 2's `raw_input()` function. That means code samples in this tutorial that use **input**`()` may not work properly in Python 2.

What is the difference between the two? In Python 2, `raw_input()` returns a string, whereas **input**`()` attempts to evaluate whatever the user types as a Python statement. Suppose a script has the following statements:

```
input("Type 33/3 and hit <Enter>: ")
raw_input("Type 33/3 and hit <Enter>: ")     # Raises NameError in Python 3
```

In Python 2, if the user follows instructions the first statement will return the integer 11.[1] The second statement will return the string `'33/3'`. In Python 3, the first statement will return the string `'33/3'`, while the second statement will result in an error because there is no function named `raw_input()`.

The exercises in this tutorial assume that **input**`()` returns a string. They may run without modification, but to run the code samples as intended in Python 2, you will need to either replace every instance of **input**`()` with `raw_input()`, or redefine the **input** function:

```
input = raw_input
```

The **input** function is not available in the `__future__` module at this time.

[1] Users who do *not* follow instructions are the primary reason this function was eliminated in Python 3.

D.4 MORE ASSISTANCE

A script named `2to3` automatically generates Python 3 code from programs written using Python 2. Though it is not necessary for this tutorial, `2to3` may be useful in making your other projects compatible with Python 3. See `docs.python.org/2/library/2to3.html` .

It is also possible to install and run multiple versions of Python—2 or 3—on the same machine using "environments" with the `conda` tool described in Section A.1.2.

APPENDIX E

Under the Hood

This appendix describes how Python handles variables and objects internally. It is not essential to understand this material to complete the exercises in this tutorial, but it may be useful when analyzing errors and writing (or reading) more advanced code in the future.

E.1 ASSIGNMENT STATEMENTS

Section 2.1 claimed that in Python, everything is an object, endowed with attributes and methods. When executing an assignment like `x=np.arange(10)`, Python binds a variable name to an object.

In some programming languages, a statement corresponding to `x=np.arange(10)` would allocate a block of memory, assign it the name `x`, then place an array of integers inside that block. The block of memory is permanently linked to the name `x`, so the name and the object it represents are essentially the same thing. This is not true in Python. Python creates an `ndarray` object, then stores its memory address in `x`. Thus, the variable `x` *points to* the `ndarray` object. The name `x` can be used to access all of the methods and attributes of the `ndarray` object, but the `ndarray` object and the variable `x` are separate entities.

One consequence of this arrangement is that two variables can point to the same object (the same memory address). Try this:

```
x = np.zeros(10)
y = x
x[1] = 1
y[0] = 1
print("x={}\ny={}".format(x, y))
```

You'll we see that `x` and `y` both point to the same array. Assignment statements can also bind an existing variable to a different object. Try the following lines of code:

```
x = np.zeros(10)
y = x
y = y + 1
```

After the first two lines, `x` and `y` point to the same array. The assignment in the third line creates a new array to hold the result of `y+1`, and then binds the variable `y` to the new array. It has no effect on `x` or the original array. (You can verify this in the Variable Explorer.)

> *An assignment statement like* `y=x` *does not permanently link* `x` *and* `y`*.*
> *They only point to the same object until one of them is reassigned.*

A consequence of this rule is that an assignment statement like `y=x+1` does not link `x` and `y` at all: `y` points to a new object created after evaluating the expression `x+1`.

If y=x binds both objects to the same memory address, how can you make an independent copy of an object? The answer depends on the type of object. You can copy a Python list by using the slicing operation:

```
x = [1, 1, 1]
y = x[:]
x == y
```

The variables x and y now contain the same elements, but are they the same list? Find out:

```
x[1] = 0
print("x={}\ny={}".format(x, y))
```

You *cannot* copy a NumPy array by using the slicing operation, however. Slicing creates a new **view** of the existing array data, not a copy.[1] Thus, modifying an element of the array also modifies any slice containing that element. Likewise, modifying a slice will change the original array. To create an independent copy of an array, you must use a method called **copy**. To understand the difference between a slice and a copy, try this:

```
w = np.zeros(10)
x = w
y = x[:]
z = y.copy()
y[0] = 1                                        # Modify a slice.
z[1] = 1                                        # Modify a copy.
print("w={}\nx={}\ny={}\nz={}".format(w, x, y, z))
```

The relation between variables and objects and the effects of assignment statements are illustrated in Figure E.1.

E.2 MEMORY MANAGEMENT

The objects created by Python do not persist forever. Python has a *garbage collection* routine that periodically checks whether *any* variables point to each object for which it has allocated memory. If no variables are bound to an object, then that object is deleted and its memory is reclaimed.

You can forcibly remove a variable by using the **del** statement: `del x`. However, Python will not destroy the object x was pointing to unless *no* variables point to that object. Memory management is under Python's control, not yours. In general, this is a good thing.

E.3 FUNCTIONS

When a variable is passed to a function, what exactly is passed?

In computer science, there are two common paradigms: *pass by value* and *pass by reference*. Pass by value means a function receives a copy of an object passed as an argument. Since the function acts on a copy, *nothing* it does to its argument affects the original object. In contrast, pass by reference means a function receives the memory address of an object passed as an argument. Since the function

[1] The array methods **ravel** and **reshape** mentioned in Section 2.2.9 also return a new view of an existing array.

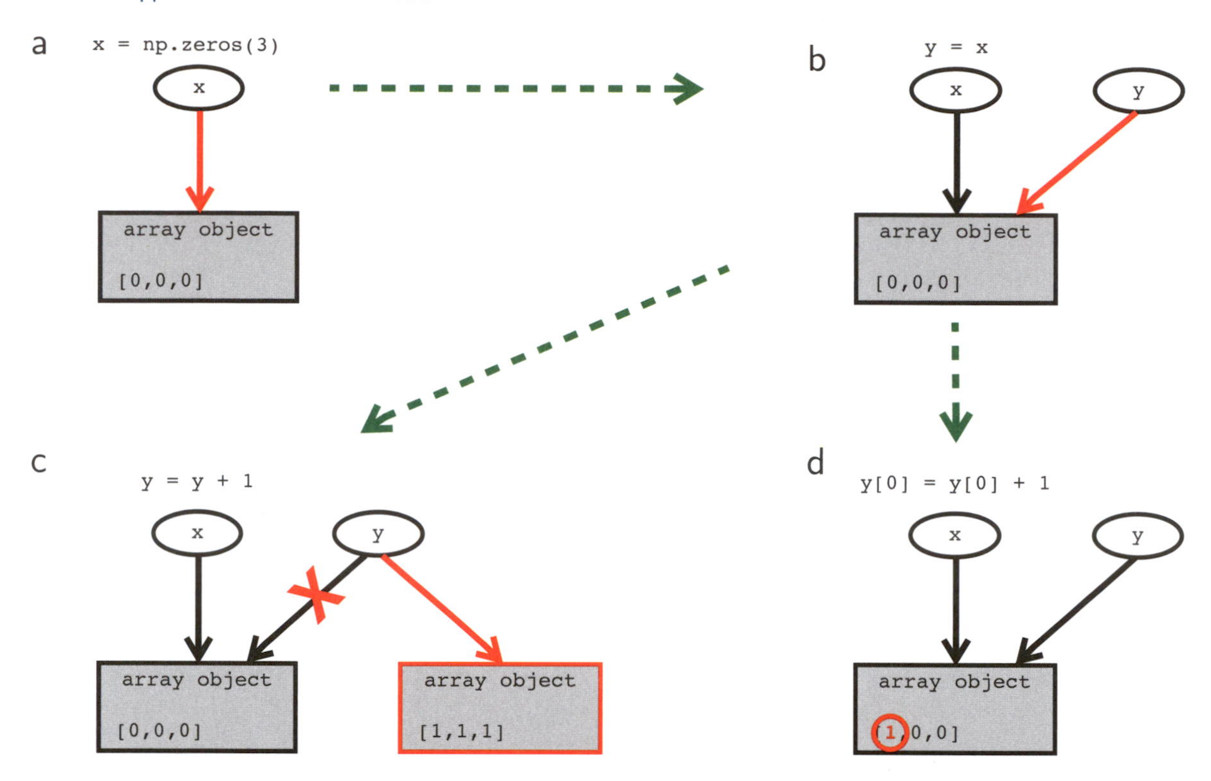

Figure E.1: Relationship between variables and objects in Python. The effects of the assignment statements are shown in red. (a) *Assigning a variable to a new object.* The call to **np.zeros**(3) creates an array object. Python then creates a variable named x that points to the array. (b) *Assigning a variable to an existing object.* The assignment statement y=x does not create a copy of x. Instead, Python binds the variable y to the *same* array as x. (c) *Reassignment of a variable.* After (b), both variables point to the same array. In the assignment y=y+1, Python creates a new array when it evaluates y+1. It then binds y to this new array, with no effect on x. (d) *Changing an element of an array.* The statement y[0]=y[0]+1 changes the value stored at position 0 of the array that y points to. Because x and y still point to the same object, x[0] also changes.

acts on the original object, *everything* it does to its argument affects the original object. If f is pass by value, then f(x) has no effect on x. If f is pass by reference, then f(x) can affect x.

Consider the following example:

```
def object_plus_one(y):
    y = y + 1

def elements_plus_one(y):
    for i in range(len(y)):
        y[i] = y[i] + 1

x = np.ones(10)
object_plus_one(x); print(x)
elements_plus_one(x); print(x)
```

One function changes x; the other does not. It appears that Python uses both paradigms, even though we did not specify either when writing the functions! How can you predict what will happen when you write a function? The key is to remember that a variable *points to* an object, but it is *not* the object. Technically, arguments to functions are always passed by value in Python. However, the "value" being passed is a reference to an object, an address in memory. Thus, a function receives a *copy* of the memory address where an argument's object is stored. It creates a new local variable for each argument and binds it to the same object as the corresponding argument. A function `f(x)` cannot modify this address and make `x` point to a *different* object, but it can use this address to access methods of the object stored there, and thereby modify the object that `x` points to. To distinguish this behavior from the ordinary usage of pass by value and pass by reference, the paradigm used in Python is often referred to as *reference passed by value,* or *pass by address.*

To better understand the behavior of functions, we next need to explore how Python finds an object when it encounters a variable name.

E.4 SCOPE

The way in which Python looks up the value of a variable is subtle but important. Python keeps track of variables using namespaces. A **namespace** is like a directory of names and objects that tells Python where to find the object associated with a variable name. The subtle point is that Python maintains multiple namespaces, each with its own **scope**. A scope is a portion of a program, like a function body or a module, in which a namespace is accessible. For example, there is a namespace that keeps track of all the variables you define at the command prompt. Its scope is *global*: Any script you run or commands you type at the command prompt can find and use these variables.

Each time Python executes a function, it creates a *local* namespace containing all variables created within the function. They are concealed from everything outside because the scope of the local namespace is limited to the function itself. Recall the `taxicab` function of the `measurements.py` module in Section 6.1:

```
# excerpt from measurements.py
def taxicab(pointA, pointB):
    """
    Taxicab metric for computing distance between points A and B.
        pointA = (x1, y1)
        pointB = (x2, y2)
    Returns |x2-x1| + |y2-y1|. Distances are measured in city blocks.
    """
    interval = abs(pointB[0] - pointA[0]) + abs(pointB[1] - pointA[1])
    return interval
```

There is no variable called `interval` in the Variable Explorer before or after you execute the `taxicab` function. It only exists in the local namespace of the function.

When you refer to a variable from within a function, Python will look for an object with that name in an expanding search. Suppose that you accidentally typed `x1` instead of `pointA[0]` in the body of `taxicab`. When you call the function, Python will search for the name `x1` in the following namespaces, in order:

Local: First, Python determines whether `x1` is defined in the function body or in its argument list. If so, it ends the search and uses the object bound to that variable. There is no variable called `x1` in this example, so Python expands the search.

Enclosing: Next, Python will determine whether the current function is defined as part of another function. If so, it will determine whether `x1` was defined within or passed as an argument to this or any other *enclosing* functions. There may be multiple nested functions, each with its own namespace for Python to search.

Global: If it still has not found `x1`, Python will check to see whether the module in which the function is defined contains a variable called `x1`. (For example, if you had added the line `x1=10` outside of any function definitions in the `measurements.py` module, Python would stop searching and use this value. If you defined `taxicab` at the command prompt or by running a script instead of importing the module, then Python would search among all the variables in your current session.)

Built-in: As a last resort, Python will check its built-in functions and parameters—those listed in `dir(__builtin__)` . If it cannot find `x1` here, it gives up and raises a `NameError` exception.

Knowing this hierarchy of $L \to E \to G \to B$ can be useful in debugging. Python may be finding a value for a variable somewhere you never intended.

The following function demonstrates Python's rules of scope.

```
# scope.py
def illustrate_scope():
    s_enclosing = 'E'
    def display():
        s_local = 'L'
        print("Local --> {}".format(s_local))
        print("Enclosing --> {}".format(s_enclosing))
        print("Global --> {}".format(s_global))
        print("Built-in --> {}".format(abs))
    display()

s_global = 'G'
illustrate_scope()
```

This script defines three variables whose names indicate their namespaces with respect to the `display()` function. When `illustrate_scope()` is called in the final line, Python executes the function body. It defines a variable in the current namespace, defines the `display()` function, and then calls this newly defined function. When Python executes `display()`, it must search for four different names. It finds `s_local` within the local namespace, but none of the other variables are defined there. Python finds `s_enclosing` in the enclosing namespace, `s_global` in the global namespace, and **abs** within Python's collection of built-in functions.

E.4.1 Name collisions

Scope is important in determining the effect of a function call. In the preceding example, all of the variables had unique names. But what happens when variables in different namespaces have the same name? Try the following example to see how Python resolves name collisions.

```
# name_collision.py
def name_collisions():
    x, y = 'E', 'E'
    def display():
        x = 'L'
        print("Inside display() ...")
        print("x= {}\ny= {}\nz= {}".format(x, y, z))
    display()
    print("Inside name_collision() ...")
    print("x= {}\ny= {}\nz= {}".format(x, y, z))

x, y, z = 'G', 'G', 'G'
name_collisions()
print("Outside function ...")
print("x= {}\ny= {}\nz= {}".format(x, y, z))
```

Each assignment statement creates a variable in the current namespace, but it has no effect on the variables in other namespaces. Thus, in this example there are three variables with the name `x`, two with the name `y`, and one with the name `z`. When Python executes `name_collision()` and `display()`, it must determine a value for each variable. It starts searching in the innermost namespace, works its way through the $L \to E \to G \to B$ hierarchy, and uses the value from the first namespace in which it locates each variable name.

Python uses namespaces, each with its own scope, to protect programs from unintended name collisions. This means that you do not have to worry about giving each variable a globally unique name when writing your own functions. However, Python cannot help you if you use the same variable name for two different purposes in the same namespace, as discussed in Section 1.4.3. Reassigning a name destroys any previous connection to another object. (Python's namespaces do not include the mind of the programmer!) Modular programming—breaking complex programs into a series of simple functions that each accomplish one thing—and descriptive variable names are the best ways to prevent this type of "name collision."

E.4.2 Variables passed as arguments

When a variable is passed to a function as an argument, Python creates a new variable in the function's local namespace and binds it to the same object as the argument. Despite referring to the same object, these two variables are independent—even if they have the *same* name. They exist in different namespaces.

If an assignment statement within a function binds an argument to a new value, there is no effect on the global variable passed as that argument. Python simply assigns the local variable to a different object. In this way, Python allows functions to *access* external variables, not *modify* them. Such modularity enforces good coding practice; however, there is an important exception to this principle. A function *can* modify an object using the *methods* of that object. The more precise rule is, "A function cannot *reassign* an external variable." If a function modifies an argument using a method of that argument, it is modifying an object that is bound to variables in multiple namespaces. This will result in side effects—whether by design or mistake.

We can now explain the behavior of `object_plus_one(x)` and `elements_plus_one(x)` in Section E.3. When `x` is passed as an argument to the function, a new local variable `y` is created in the function's local namespace and bound to the same array as `x`, as in Figure E.1b. The assignment `y=y+1` binds the local `y` to a new object with no effect on the global `x`, as in Figure E.1c. Because no value is returned by the function, and Python deletes local variables after evaluating the function, `object_plus_one(x)` has no external effects—it accomplishes nothing.

In contrast, `elements_plus_one(x)` does not reassign its local variable `y` to a new object. It uses an array method to modify the data of `y`. (See Section 2.2.6, page 24.) The statement `y[0]=y[0]+1` changes the value of the first element of the array, as shown in Figure E.1d. Because the variable `y` in the function's local namespace and the variable `x` in the global namespace point to the *same* array, `x[0]` is also changed. This continues as the function iterates over the loop. Unlike `object_plus_one(x)`, `elements_plus_one(x)` produces a side effect: It changes the value of `x`.

E.5 SUMMARY

When we think about an assignment statement like `x=2`, it's natural to imagine that some memory is set aside for `x`, then the value 2 is stored there. Python does not use this approach. Instead, it creates an integer object with the value 2, then binds the variable `x` to that object.

> *In Python, objects and variables exist independently of one another. After an assignment statement, a variable points to the address in memory where an object is stored.*

Variables have a scope: They can only be accessed by certain portions of a program, and Python uses namespaces to prevent conflicts between variables with the same name. There is no such protection for the objects they are bound to. Multiple variables can point to the same object, and this can lead to unexpected results when you are trying to carry out different operations on what you thought were copies of the same data. Functions create temporary variables that are initially bound to the same objects as their arguments. This, too, can lead to unexpected behavior.

Errors due to improper use of Python's system of variables bound to objects can be difficult to diagnose. The best way to avoid them is to plan carefully when writing code. When you make an assignment or write a function, consider: Do I want two variables that point to the *same* data? Or do I want to make a *copy* of the data?

APPENDIX F

Answers to "Your Turn" Questions

Your Turn 2B:

The **np.array** function is called on a list of lists—a two-element list whose elements are themselves three-element lists of integers—so it creates a 2×3 array to store the data. Python starts filling in the (0,0) cell of the array. Each comma moves to the next column in the same row of the array. At the end of the first list, that is, after "`]`", Python moves to the first column of the next row of the array. The resulting array is equivalent to a matrix with two rows and three columns containing the original data.

Your Turn 2C:

The **np.linspace** function ends exactly at 10 and uses whatever spacing it needs to accomplish that. In contrast, **np.arange** uses the exact spacing of 1.5 and ends just before it reaches 10.

Your Turn 2D:

To get the odd-index elements, use `a[1::2]`. This instructs Python to start at offset 1, then step by 2 until it reaches the end of the array.

Your Turn 2E:

Python concatenates three strings to produce the output: `s`, a single space, and `t`.

Your Turn 3A:

a.

```
x = np.linspace(-3, 3, 101)
y = np.exp(-x**2)
```

b.

```
from scipy.special import factorial
mu, N = 2, 10
n_list = np.arange(N + 1)
poisson = np.exp(-mu) * (mu**n_list) / factorial(n_list)
```

Your Turn 4A:

```
# fancy_plot.py
x_min, x_max = 0, 4
num_points = 51
x_list = np.linspace(x_min, x_max, num_points)
y_list = x_list**2
plt.plot(x_list, y_list, 'r', linewidth=3)
```

```
ax = plt.gca()
ax.set_title("My Little Plot", fontsize=16)
ax.set_xlabel("$x$", fontsize=24)
ax.set_ylabel("$y = x^2$", fontsize=24)
```

Note that we have to set the font size for each element of the figure separately.

Your Turn 4B:

```
# legend.py
ax = plt.gca()
ax.legend( ("sin(2$\\pi$x)",\
            "sin(4$\\pi$x)",\
            "sin(6$\\pi$x)") )
```

Your Turn 6A:

The function should be similar to taxicab with a modified distance function. The docstring should specify that each point contain three elements.

```
# measurements.py
def crow(pointA, pointB):
    """
    Distance between points A and B "as the crow flies."
        pointA = (x1, y1, z1)
        pointB = (x2, y2, z2)
    Returns sqrt((x2-x1)**2 + (y2-y1)**2 + (z2-z1)**2)
    """
    distance = np.sqrt( (pointA[0] - pointB[0])**2 + \
                        (pointA[1] - pointB[1])**2 + \
                        (pointA[2] - pointB[2])**2 )
    return distance
```

Your Turn 6B:

a. `x_step = 2*x_step - 1`

b. `x_position = np.cumsum(x_step)` or `x_position = x_step.cumsum()`

c.

```
# random_walk.py
from numpy.random import random as rng
num_steps = 500
x_step = rng(num_steps) > 0.5
y_step = rng(num_steps) > 0.5
x_step = 2*x_step - 1
y_step = 2*y_step - 1
x_position = np.cumsum(x_step)
y_position = np.cumsum(y_step)
plt.plot(x_position, y_position)
plt.axis('equal')
```

Your Turn 6C:

By default, `plt.hist` and `np.histogram` set up ten equally spaced bins. The variable `bin_edges` is an array with *eleven* elements—the *edges* of the bins. The first element is the smallest entry in the data list; the last element is the largest entry in the data list. Thus, each bin's width is `(data.max()-data.min())/10`. Meanwhile, `counts[i]` contains all the elements of `data` that fall between `bin_edges[i]` and `bin_edges[i+1]`. Thus, `bin_edges` will always contain one more element than `counts`.

Your Turn 6D:

```
# surface.py
from mpl_toolkits.mplot3d import Axes3D
points = np.linspace(-1, 1, 101)
X, Y = np.meshgrid(points, points)
Z = X**2 + Y**2
ax = Axes3D(plt.figure())
ax.plot_surface(X, Y, Z)
# ax.plot_surface(X, Y, Z, rstride=1, cstride=1)
```

The commented line uses all points in the mesh. Use it for a coarse grid with fewer points.

Your Turn 6F:

a.

```
from scipy.integrate import quad
def f(x): return x**2
integral, error = quad(f, 0, 2)
print(integral - 2**3 / 3)
```

b.

```
# quadrature.py
from scipy.integrate import quad
xmax = np.linspace(0, 5, 51)
integral = np.zeros(xmax.size)
def integrand(x): return np.exp(-x**2/2)
for i in range(xmax.size):
    integral[i], error = quad(integrand, 0, xmax[i])
plt.plot(xmax, integral)
```

Your Turn 6G:

Driving the system near its resonant frequency ($\omega_0 = 1$) yields a strong response. The driven mode of frequency $\omega = 0.8$ is superposed with the natural modes required to satisfy the initial conditions. Interference between modes of different frequencies produces a pattern of beats.

Your Turn 6H:

The vector at the point (x, y) has components $(y, -x)$. This arrow is always perpendicular to the vector from the origin to its base point, just like the velocity vector field of a rigid, spinning disk.

Your Turn 6I:

a. This example gives trajectories that all spiral into the origin, because we added a small component to $\boldsymbol{V}$ that points radially inward.

b. This example gives a fixed point at the origin of "saddle" type. One of the initial conditions runs smack into the fixed point at the origin, but the others veer away and run off to infinity.

Your Turn 8A:

The resulting image is a negative of the original with no shades of gray, just black and white. Everything darker than the mean lightness became fully white, and everything lighter than the mean became fully black. In performing the array operation, NumPy has modified the data type of the array from `uint8` to `bool`. PyPlot has no trouble generating images and saving files with arrays of different data types, but the array is no longer a collection of 8-bit luminance values.

Your Turn 9A:

a. In the Python indexing scheme, the only allowed index for F is $(k,\ell) = (0,0)$. Thus, $C_{i,j} = F_{0,0} \cdot I_{i,j} = I_{i,j}$.

b. Again using the Python indexing scheme, the allowed indices in Equation 9.1 run from $(i,j) = (0,0)$ with $(k,\ell) = (0,0)$ to $(i,j) = (M+m-2, N+n-2) = ([M-1]+[m-1], [N-1]+[n-1])$ with $(k,\ell) = (m-1, n-1)$. Thus, C contains $(M+m-1) \times (N+n-1)$ entries.

Acknowledgments

Many students and colleagues taught us tricks that ended up in this tutorial, especially Tom Dodson. Alexander Alemi, Steve Baylor, Gary Bernstein, Kevin Chen, R. Michael Jarvis, Michael Lerner, Dave Pine, and Jim Sethna gave expert advice and caught many errors. Kerry Watson at Anaconda, Inc. offered clarifications on the material in Appendix A. Three anonymous reviewers not only gave us expert suggestions, but also questioned sharply the goals of the book in ways that helped us to clarify those goals. Nily Dan reminded us to strive for utility, not erudition. Kim Kinder provided encouragement and diversion in equal measure, at just the right times, throughout the writing of this book.

This tutorial was set in LaTeX using the `listings` package. Jobst Hoffman and Heiko Oberdiek kindly supplied personal help. (The package settings that generated our code listings are available at the book's blog.)

We're grateful to Ingrid Gnerlich and Mark Bellis at Princeton University Press for their meticulous care with this unusual project, and to Teresa Wilson and Terry Kornak for their exacting, and insightful, copyediting.

This tutorial is partially based on work supported by the United States National Science Foundation under Grants EF–0928048, DMR–0832802, and (updated edition) PHY–1601894. The Aspen Center for Physics, which is supported by NSF grant PHYS–1066293, also helped to bring it to completion. Any opinions, findings, conclusions, jokes, or recommendations expressed in this book are those of the authors and do not necessarily reflect the views of the National Science Foundation.

References

If you can know everything about anything, it is not worth knowing.
—Robertson Davies

Starting from the foundation in this tutorial, you may be able to get the advanced material you need for a particular problem just from web searches. Some other references that we found helpful appear below.

As your skills develop, you may start writing longer and more complex codes. Then it will pay off to make an additional investment in learning about basic software engineering practices (for example, in Scopatz & Huff, 2015) and specifically about user-defined classes (for example, in Downey, 2012; Guttag, 2016; Scopatz & Huff, 2015). The PEP 8 Style guide for Python also provides guidelines for writing Python code in a standard, readable way: `www.python.org/dev/peps/pep-0008/`.

This book has alluded to the free LaTeX typesetting system, in the context of graph labels and Jupyter notebooks. You can obtain it, and its documentation, from `www.latex-project.org/get/`.

A blog accompanies this book. It can be accessed via `press.princeton.edu/titles/11349.html`, or directly at `physicalmodelingwithpython.blogspot.com/`. Here you will find data sets, code samples, errata, additional resources, and extended discussions of the topics introduced in this book.

A counterpart to this tutorial covers similar techniques, but with the MATLAB programming language (Nelson, 2015).

BERENDSEN, H J C. 2011. *A student's guide to data and error analysis.* Cambridge UK: Cambridge Univ. Press.

CROMEY, D W. 2010. Avoiding twisted pixels: Ethical guidelines for the appropriate use and manipulation of scientific digital images. *Sci. Eng. Ethics*, **16**(4), 639–667.

DOWNEY, A. 2012. *Think Python.* Sebastopol CA: O'Reilly Media Inc. `www.greenteapress.com/thinkpython/thinkpython.pdf`.

GUTTAG, J V. 2016. *Introduction to computation and programming using Python.* 2d ed. Cambridge MA: MIT Press.

HILL, C. 2015. *Learning scientific programming with Python.* Cambridge UK: Cambridge Univ. Press.

IVEZIC, Ž, CONNOLLY, A J, VANDERPLAS, J T, & GRAY, A. 2014. *Statistics, data mining, and machine learning in astronomy: A practical Python guide for the analysis of survey data.* Princeton NJ: Princeton Univ. Press.

LANDAU, R H, PÁEZ, M J, & BORDEIANU, C C. 2015. *Computational physics: Problem solving with computers.* 3rd ed. New York: Wiley-VCH. `physics.oregonstate.edu/~rubin/Books/CPbook/index.html`.

LANGTANGEN, H P. 2016. *A primer on scientific programming with Python.* 5th ed. Berlin: Springer.

LIBESKIND-HADAS, R, & BUSH, E. 2014. *Computing for biologists: Python programming and principles.* Cambridge UK: Cambridge Univ. Press.

LUTZ, M. 2014. *Python pocket reference.* 5th ed. Sebastopol CA: O'Reilly Media Inc.

NELSON, P. 2015. *Physical models of living systems.* New York: W. H. Freeman and Co.

NELSON, P. 2017. *From photon to neuron: Light, imaging, vision.* Princeton NJ: Princeton Univ. Press.

NEWMAN, M. 2013. *Computational physics.* Rev. and expanded ed. CreateSpace Publishing.

PÉREZ, F, & GRANGER, B E. 2007. IPython: A system for interactive scientific computing. *Computing in Science and Engineering*, **9**(3), 21–29.

PINE, D. 2014. *Introduction to Python for science.* `github.com/djpine/pyman`.

ROUGIER, N P, DROETTBOOM, M, & BOURNE, P E. 2014. Ten simple rules for better figures. *PLoS Comput. Biol.*, **10**(9), e1003833.
`journals.plos.org/ploscompbiol/article?id=10.1371/journal.pcbi.1003833`.

SCOPATZ, A, & HUFF, K D. 2015. *Effective computation in physics.* Sebastopol CA: O'Reilly Media.

SHAW, Z. 2017. *Learn Python 3 the hard way: A very simple introduction to the terrifyingly beautiful world of computers and code.* Upper Saddle River NJ: Addison-Wesley.

WANG, J. 2016. *Computational modeling and visualization of physical systems with Python.* Hoboken NJ: Wiley.

ZEMEL, A, REHFELDT, F, BROWN, A E X, DISCHER, D E, & SAFRAN, S A. 2010. Optimal matrix rigidity for stress-fibre polarization in stem cells. *Nature Physics*, **6**(6), 468–473.

Index

Bold references indicate the main or defining instance of a key term.